Disclaimer

The publisher of this book is by no way associated with the National Institute of Standards and Technology (NIST). The NIST did not publish this book. It was published by 50 page publications under the public domain license.

50 Page Publications.

Book Title: Impact of Sprinklers on the Fire Hazard in Dormitories: Sleeping Room Fire Experiments.

Book Author: Daniel Madrzykowski; William D. Walton;

Book Abstract: The objective of this study was to compare the levels of hazard created by room fires in a dormitory building with and without automatic fire sprinklers in the room of fire origin. This report describes a series of experiments where fires were initiated in a dormitory sleeping room. The description of the experimental conditions includes: the geometry and construction of the building, the fuel load in the sleeping rooms, and the location of the instrumentation used to measure gas temperature, oxygen, carbon dioxide and carbon monoxide concentrations and heat flux. Smoke alarm activation and sprinkler activation times are also reported. Five experiments were conducted. In two of the experiments, the door between the sleeping room (room of fire origin) and the corridor was closed. In the other three experiments the door from the sleeping room (room of fire origin) remained open to the corridor. In each case, door closed or door open, one of the experiments was sprinklered. The results from the experiments comparing the sprinklered and non-sprinklered sleeping room are presented. The results from these experiments demonstrate the potential life safety benefits of smoke alarms, compartmentation, and automatic fire sprinkler systems in college dormitories and similar occupancies. These experiments were conducted by NIST in cooperation with the University of Arkansas and the Fayetteville Fire Department.

Citation: NIST TN - 1658

Keywords: corridor tests; dormitories; fire data; gas concentrations; heat flux; large scale fire tests; sprinklers; temperature measurements

NIST TN 1658

Impact of Sprinklers on the Fire Hazard in Dormitories: Sleeping Room Fire Experiments

Daniel Madrzykowski
William D. Walton

Sponsored in part by
Department of Homeland Security
Federal Emergency Management Agency
United States Fire Administration

National Institute of Standards and Technology
Technology Administration, U.S. Department of Commerce

NIST TN 1658

Impact of Sprinklers on the Fire Hazard in Dormitories: Sleeping Room Fire Experiments

Daniel Madrzykowski
William D. Walton

Fire Research Division
Building and Fire Research Laboratory
National Institute of Standards and Technology
Gaithersburg, MD 20899-8661

January 2010

Department of Homeland Security
Janet Napolitano, *Secretary*
Federal Emergency Management Association
Craig Fugate, *Administrator*
United States Fire Administration
Kelvin J. Cochran, *Assistant Administrator*

U.S. Department of Commerce
Gary Locke, *Secretary*
National Institute of Standards and Technology
Patrick Gallagher, *Director*

Impact of Sprinklers on the Fire Hazard in Dormitories: Sleeping Room Fire Experiments

Daniel Madrzykowski and William D. Walton
Building and Fire Research Laboratory
National Institute of Standards and Technology

Abstract

As part of a U.S. Fire Administration (USFA) initiative to improve fire safety in college housing, the National Institute of Standards and Technology (NIST) conducted two series of full-scale fire experiments in abandoned dormitory buildings. The objective of the study was to compare the levels of hazard created by room fires in a dormitory building with and without automatic fire sprinklers in the room of fire origin.

Five experiments were conducted which included variables such as an open/closed door between the room of origin and the corridor, and with/without sprinklers in the room of origin. Gas temperatures and concentrations (oxygen, carbon monoxide and carbon dioxide) were measured continuously in each experiment. This report contains analysis of the data collected and a detailed discussion of the experimental conditions, such as fuel load in the room of origin, geometry and construction of the room, and the locations of instrumentation. The results of this study demonstrate the potential life safety benefits of smoke alarms, compartmentation, and automatic fire sprinkler systems in college dormitories and similar occupancies. These experiments were conducted by NIST in cooperation with the University of Arkansas and the Fayetteville Fire Department.

The other series of experiments was conducted with the fires initiated in a day room area open to the corridor of the dormitory. These experiments were conducted by NIST in cooperation with the Myrtle Beach Air Force Base Redevelopment Authority, the Myrtle Beach Fire Department, and the Bureau of Alcohol, Tobacco and Firearms (ATF). The results of these experiments are presented in the report, *Impact of Sprinklers on the Fire Hazard in Dormitories: Day Room Fire Experiments, NISTIR 7120.*

For further information on the USFA College Campus Fire Safety Program contact: www.usfa.dhs.gov/citizens/college/.

Key Words: corridor tests; dormitories; fire data; gas concentrations; heat flux; large scale fire tests; sprinklers; temperature measurements; tenability

Disclaimer

Certain trade names or company products are mentioned in the text to specify adequately the experimental procedure and equipment used. In no case does such identification imply recommendation or endorsement by the National Institute of Standards and Technology, nor does it imply that the equipment is the best available for the purpose.

Regarding Non-Metric Units: The policy of the National Institute of Standards and Technology is to use metric units in all its published materials. To aid the understanding of this report, in most cases, measurements are reported in both metric and U.S. customary units, in some cases the conversion will have been rounded.

Acknowledgements

The authors would like to thank the University of Arkansas, especially Mr. John Wichser, Associate Director for Residential Facilities and the Fayetteville Fire Department, especially Chief Terry Lawson, for supporting the conduct of the experiments and providing logistical and fire fighting support. The authors would like to recognize Mr. Laurean DeLauter and Mr. Jay McElroy of the Building and Fire Research Laboratory at NIST, Ms. Sarah Thompson and Mr. William Twilley formerly with NIST and Mr. Scott Dillon formerly of the ATF Fire Research Laboratory for the invaluable support that they provided during this series of experiments.

The authors also wish to thank Jonathan Kent for the application of his programming talents which enabled improved data analysis capabilities and graph generation.

Finally, the authors express gratitude to the United States Fire Administration and Mr. Larry Maruskin for their sponsorship and support of this important research program.

Table of Contents

vii

List of Figures

x

List of Tables

1 Introduction

From 2003 through 2006 there was an estimated annual average of 3,370 structure fires in dormitory or barracks occupancies in the United States. Of the reported structure fires in dormitory properties, only 5% of the fires originated in the bedroom or sleeping room. However, this small percentage of fires resulted in 62% of the civilian fire deaths and 26% of civilian fire injuries [1].

As part of the U.S. Fire Administration (USFA) initiative to improve fire safety in college housing, the National Institute of Standards and Technology (NIST) conducted two series of full-scale fire experiments in abandoned dormitory buildings. The objective of the study was to compare the levels of hazard created by room fires in a dormitory building with and without automatic fire sprinklers in the room of fire origin. In addition, the effect of a closed door was examined.

One series of experiments was designed with the fires starting in a day room or lounge area open to the corridor of the dormitory. These experiments were conducted by NIST in cooperation with the Myrtle Beach Air Force Base Redevelopment Authority, the Myrtle Beach Fire Department, and the Bureau of Alcohol, Tobacco and Firearms (ATF). The results of these experiments are presented in the report, *Impact of Sprinklers on the Fire Hazard in Dormitories: Day Room Fire Experiments, NISTIR 7120* [2].

The series of experiments presented in this report were designed with the fires starting in a dormitory sleeping room. These experiments were conducted by NIST in cooperation with the University of Arkansas and the Fayetteville, Arkansas Fire Department.

This paper documents experiments conducted to examine the fire development, the spread of hot gases outside the fire room, and the effectiveness of an automatic sprinkler system in suppressing the fire. Five fire experiments were conducted:

        1) Fire room door closed, non-sprinklered
        2) Fire room door closed, sprinklered
        3) Fire room door open, sprinklered
        4) Fire room door open, non-sprinklered
        5) Fire room door open, non- sprinklered

The experiments were conducted in an abandoned college dormitory building located on the University of Arkansas Campus in Fayetteville, Arkansas.

2   Technical Approach

Opportunities to conduct real scale fire experiments in a structure are rare. These experiments are important, because they provide the nearest simulation of "real world" conditions, in terms of geometry, materials and building elements, heat loss to the structure, ventilation, and volume. Fire experiments were conducted in an unoccupied college dormitory building that was slated for demolition.

Measurements were made to quantify and differentiate the level of hazard and the rate of hazard development in both sprinklered and non-sprinklered student sleeping rooms and the adjacent corridor. Temperature, oxygen, carbon dioxide and carbon monoxide concentration measurements were made in the sleeping rooms (room of fire origin) and in the corridor. Smoke alarm activation and sprinkler activation times were also recorded. In addition, heat flux measurements in the corridor were made to provide information on the thermal conditions to which fire fighters approaching the room of origin would be exposed. The experiments were also recorded with video cameras.

Each sleeping room was furnished with similar furniture, computers and related equipment, clothing and decorations representative of an actual dorm room. The fuel load was described, weighed and documented for each room, as means to obtain a similar fuel load in each dorm room. The experimental approach permitted the examination of the impact of an automatic fire sprinkler (suppression system) in a real-scale college dormitory building. The other key issue examined was the impact of a closed door (compartmentation) on conditions inside the room of fire origin as well as the spread of heat and combustion gases to the corridor. Commercially available ionization smoke alarms were also installed in the sleeping rooms and in the corridor to measure the time to alarm. Each experiment involved the ignition of a small fire in one sleeping room.

2.1   Experimental Arrangement

With the cooperation of the University of Arkansas in Fayetteville and the Fayetteville Fire Department, the experiments were conducted in an unoccupied dormitory building that was constructed in the 1950s. The building was of fire resistive construction.

2.2   Dormitory Building

The building used for the experiments was a four-story, dormitory building. The floors and ceilings of the building were poured concrete and the walls were made from concrete block with brick covering the exterior. Each floor of the building had four residential wings, two on the north side of the building and two on the south. The two wings on each of the north and the south sides of the building were separated by an open air stairway. The north and south sides of the building were joined by bathroom and shower facilities on the east and western portion of the building with an open air walkway in between. The overall footprint of the building, including the two courtyards, was approximately 95 m (310 ft) by 39 m (128 ft).

The first floor of the northwest wing of the building was chosen for the experiments based on the pristine condition of the dorm rooms. Photographs of the exterior and interior of the northwest wing are shown in Figure 2.2-1 and Figure 2.2-2. In Figure 2.2-2 the sleeping rooms are on the left side of the corridor and a courtyard can be seen through the windows on the right.

The length of the corridor was 19.81 m (65.00 ft). The corridor was 2.54 m (8.33 ft) wide with a ceiling height of 2.39 m (7.83 ft). The rooms were located on the north side of the corridor. Windows comprised the upper portion of the south wall of the corridor. The single pane windows were 1.33 m (4.38 ft) high with 0.83 m (2.70 ft) window sill height above the floor. A door, 0.98 m (3.23 ft) wide and 2.02 m (6.63 ft) high, provided access to the corridor from the east end of the corridor. This door was kept closed during the fire experiments. A smoke curtain made of gypsum board was installed on the west end of the corridor to limit smoke flow into the portion of the corridor that led to the shower rooms and connected the southwest corridor of the building. An area 1.55 m (5.08 ft) wide and 1.07 m (3.5 ft) was left open under the smoke curtain to prevent pressure increase in the corridor during the fire.

Figure 2.2-1. Photograph of the outside of the northwest wing of the dormitory building, looking southwest.

Figure 2.2-2. Photograph of the instrumented corridor in the northwest wing of the dormitory building, looking east.

Five rooms of identical size, Dorm Room 1 through Dorm Room 5, each connected to the corridor, were used for the fire experiments (Figure 2.2-3).   The inside dimensions of each room were: 4.47 m (14.66 ft) deep, 3.45 m (11.33 ft) wide and 2.39 m (7.83 ft) high. The doorway to the corridor was 0.73 m (2.40 ft) wide by 2.03 m (6.65 ft) high.

The dorm room doors were solid core, wood doors, 35.00 mm (1.375 in) thick. Gaps between the steel door frame and the door were 5 mm ± 2 mm, due in part to rubber bumpers mounted in the door frames. There was no gasket of any kind used to seal the gap between the doors and the door frames. There was a 12 mm (0.5 in) gap between the bottom of the door and the corridor floor.

The single pane windows on the exterior wall of the dorm room were composed of three sections of glass 0.84 m (2.77 ft) wide and 1.49 m (4.88 ft) high with a window sill height of 0.83 m (2.71 ft). The glass was mounted in an aluminum frame. The floor plan of a dorm room is given in Figure 2.2-4.

4

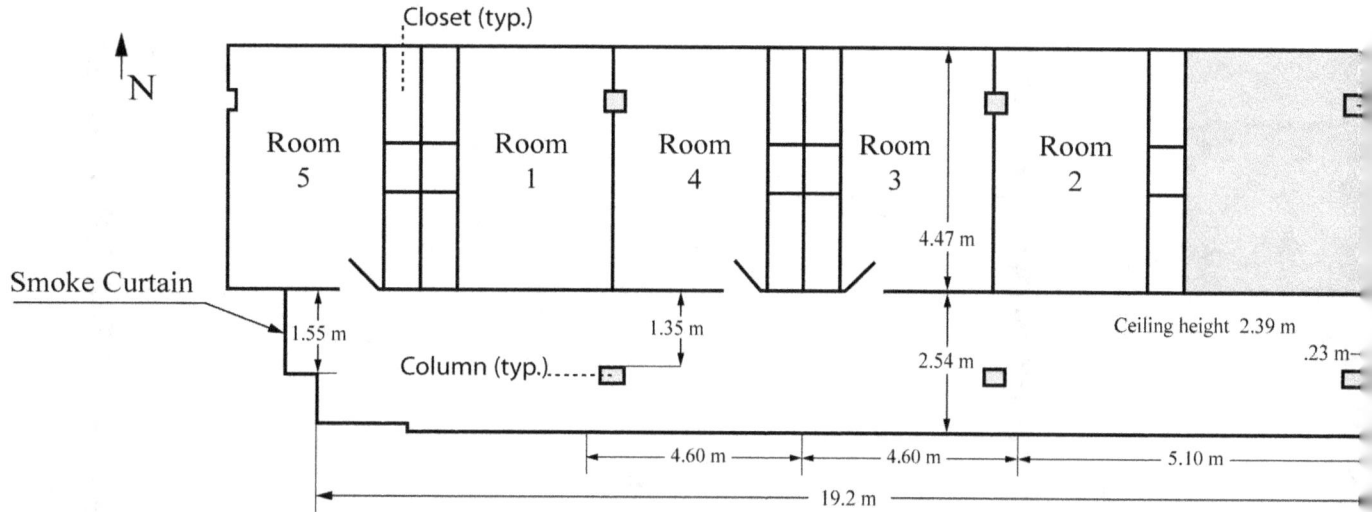

Figure 2.2-3. Floor plan of the northwest corridor

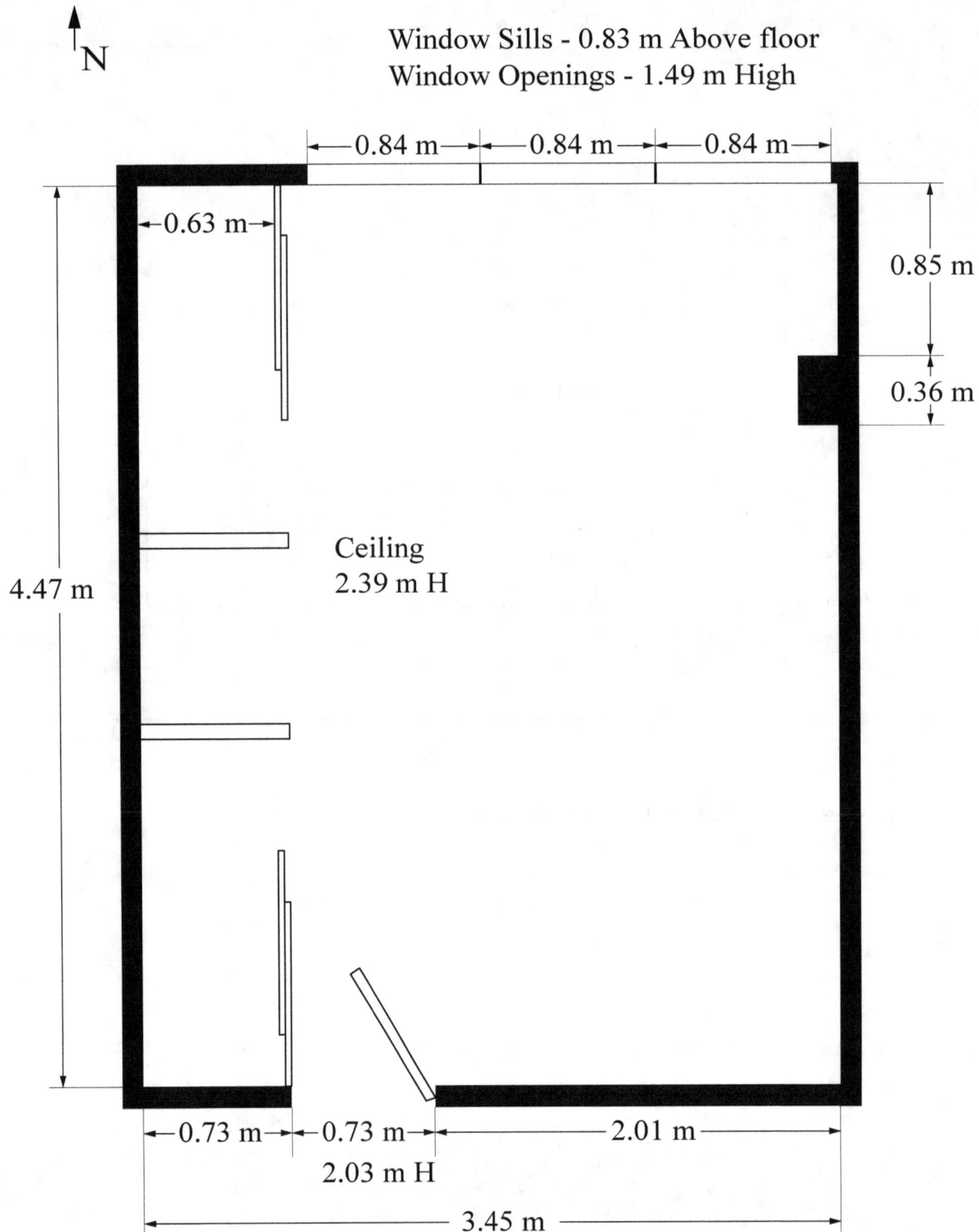

N

Window Sills - 0.83 m Above floor
Window Openings - 1.49 m High

←—0.84 m—→←—0.84 m—→←—0.84 m—→

←0.63 m→

0.85 m

0.36 m

4.47 m

Ceiling
2.39 m H

←—0.73 m—→←—0.73 m—→←————2.01 m————→

2.03 m H

←————————3.45 m————————→

Figure 2.2-4. Floor plan of a typical dorm room used in the experiments

2.3 Room Contents

Each room was furnished in a similar manner. The furnishings were typical of items found in college dormitory sleeping rooms: twin beds, desks, computers, books, clothing, towels, posters, carpeting, a television, and a radio. A complete list of the items in the rooms and a brief description and the mass of each item is given in Table 2.3-1 through Table 2.3-5. The tables are categorized by the type or area of fuel load and include the fixed furnishings and interior finish (Table 2.3-1), the bed fuel package (Table 2.3-2), the desk fuel package (Table 2.3-3), clothing and closet contents (Table 2.3-4), and the ignition fuel package (Table 2.3-5).

N

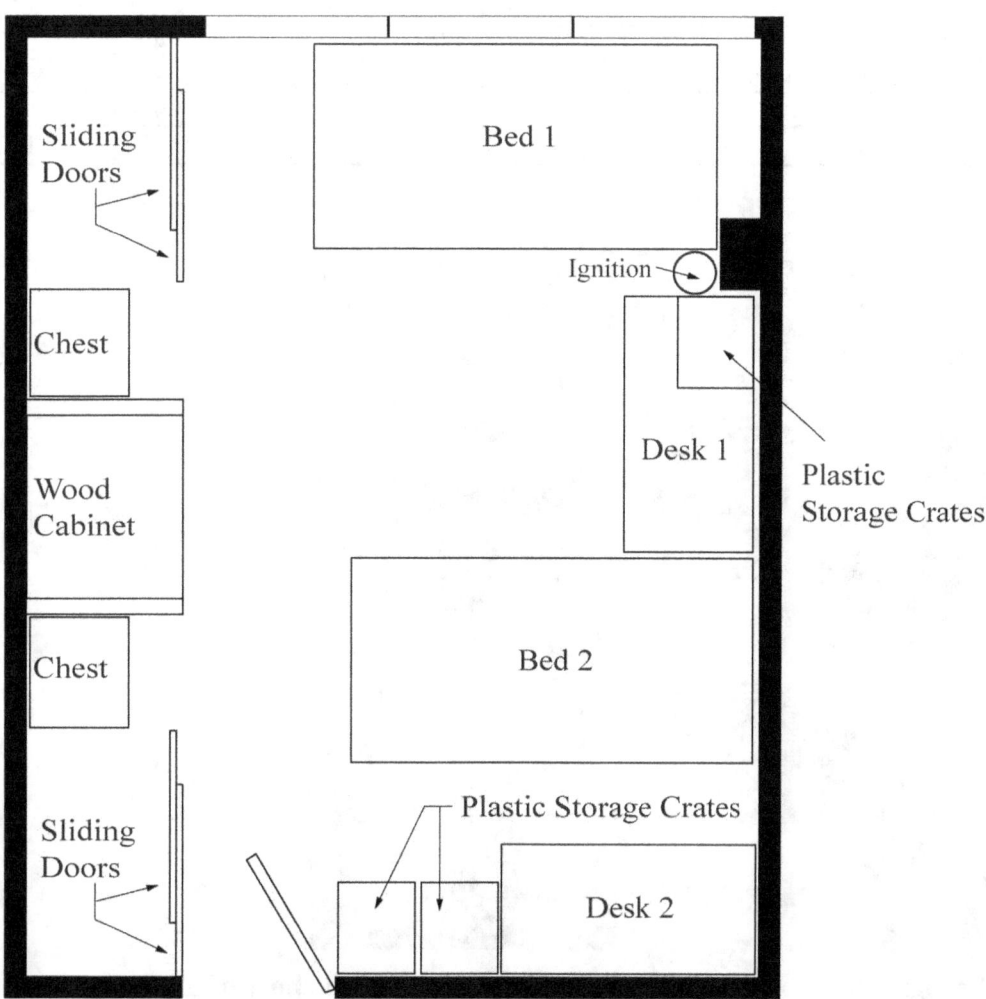

Figure 2.3-1. Furnishing arrangement of a typical dorm room used in the experiments

Table 2.3-1 contains the portion of the fuel load that was composed of interior finish materials, wall hangings and the wooden fixed furniture which lined one side of each of the dorm rooms. A photograph of the wooden chest of drawers, closet and shelves is provided in Figure 2.3-5.

Table 2.3-1. Fixed furnishings and interior finish

Item	#	Materials	Dimensions (m)			Individual Mass (kg)	Total Mass (kg)
			W (m)	L (m)	H (m)		
Padding	1	Urethane foam	12.55 (area m²)		0.012	7.1	7.1
Carpeting	1	88% olefin, 12% nylon, "26 oz", 14 tuft, 5200 density	12.55 (area m²)		0.008	25.0	25.0
Bulletin board	1	Cork Board w/ Aluminum Frame	0.610	0.914	0.013	1.75	3.5
Posters	4	2 – poster board 2 – poster paper					0.3
Chests	2	Pine	0.514	0.775	0.464	26.5	53.0
Closet doors	4	Peg board (hard board)	0.927	0.006	1.962	6.4	25.6
Center shelving	1	Pine	0.305	4.775	0.025	12.4	12.4
Upper shelf	2	Pine	0.292	1.753	0.018	6.0	12.0
Lower shelf	2	Pine	0.387	1.753	0.018	4.5	9.0
Upper slider	4	Peg board (hard board)	0.318	1.219	0.006	2.6	10.4
Roll-up blind	1	Vinyl 9.5 mm slat	2.667	1.829		4.4	4.4
Total							162.7

Figure 2.3-2. Photograph of fixed furnishings composed of wood chest of drawers, wood shelves and peg board cabinet closet doors

Table 2.3-2 lists all of the materials that made up the fuel load of the bed and items that were placed on the bed. A photograph of the bed fuel load is provided in Figure 2.3-3.

Table 2.3-2. Bed fuel package

Item	#	Materials	Dimensions			Individual Mass (kg)	Total Mass (kg)
			W (m)	L (m)	H (m)		
Mattress	2	Resin Treated Textile clippings – 58% Urethane foam – 42%	0.97	1.9	0.63	16.2	32.4
Bed Pad	2	100% polyurethane foam	0.86	1.83	0.03	0.7	1.4
Bedding, Flat Sheet, Fitted Sheet, Pillow Case	2	100% Cotton	Twin bed size			1.2	2.4
Pillows	4	Cover – 50% Cotton, 50% polyurethane Stuffed with 100% polyester fiber	0.5	0.71		0.75	3.0
Stuffed animals	5	Polyester filled					2.6
Backpacks	2	Nylon				0.4	0.8
Bed Cover	2	100% Polyester				1.3	2.6
Total							45.2

Figure 2.3-3. Photograph of the bed fuel package

9

Table 2.3-3 lists the materials that made up the desk fuel package, which included the desk chair, computer equipment, plastic crate shelving with notebooks and paper products. Figure 2.3-4 shows the desk fuel package.

Table 2.3-3. Desk fuel package

Item	#	Description / Materials	Dimensions			Individual Mass (kg)	Total Mass (kg)
			W (m)	L (m)	H (m)		
Desk Chair	2	Plastic Shell - PP Seat Cover - Olefin 40%, Acrylic 40%, Nylon 20% Pad – Polyurethane Foam 100%	0.53	0.58	0.91	10.7	21.4
Back cushion			0.46	0.41	0.08		
Seat cushion			0.53	0.46	0.09		
Desk	2	Laminated particle board	1.21	0.62	0.72	39.0	78.0
Desk lamp	2	Polystyrene	0.58	0.23	0.25	0.7	1.4
Monitor	2	15 in. monitor / FR ABS Plastic Monitor Case	0.36	0.39	0.3	12.9	25.8
Keyboard	2	Plastic	0.45	0.2	0.04	1.5	3.0
Computer	2	Plastic front, steel shell	0.4	0.38	0.1	7.8	15.6
Printer	1	Plastic case	0.48	0.46	0.22	20.2	20.2
File crates	8	Self stacking / HDPE	0.44	0.36	0.29	1.0	8.0
Stationary		Notebooks, paper etc				107.2	107.2
Telephones	2	Plastic Shell, Metal Base				0.6	1.2
Total							281.8

Figure 2.3-4. Photograph of the desk fuel package

Table 2.3-4 provides a list of clothing materials which were located in the closet, in the laundry hampers and, in the case of shoes on the floor, next to the bed. The radio, television, towels and toiletries were located on the wooden fixed furnishings as shown in Figure 8. Smaller items such as the shampoos and lotions are not included in Table 2.3-4.

Table 2.3-4. Clothing and closet contents

Item	#	Materials	Dimensions			Individual Mass (kg)	Total Mass (kg)
			W (m)	L (m)	H (m)		
Radio/CD Player	1	Plastics	0.362	0.21	0.15	2.3	2.3
Clothing	1	Cotton, polyester				90.0	90.0
Shoes (pair)	6	Fabric, plastics				0.48	2.9
Handbags	4	Fabric & vinyl				0.48	1.9
Plastic Hangers 1	5	Hard plastic, metal hooks		0.41		0.04	0.2
Plastic Hangers 2	10	Flexible round plastic		0.41		0.04	0.4
Television	1	Plastic Case	0.510	0.460	0.510	17.6	17.6
Towels	2	100% Cotton	0.762	1.38		0.5	1.0
Laundry Basket	1	Polypropylene	0.457 (dia)		0.33	0.5	0.5
Clothes Hamper	1	Polypropylene w/cover	0.406 (dia)		0.66	1.4	1.4
Plastic Storage Box	1	Polypropylene	0.5	1	0.15	2.8	2.8
Bed cover in storage box	1	100% Polyester	Twin bed size			1.3	1.3
Total							122.3

Figure 2.3-5. Photograph showing clothing and closet fuel load arrangement

11

The final fuel package, Table 2.3-5, was the waste basket and contents where ignition took place. The ignition fuel package was intended to represent a small trash container and its contents. The ignition fuel package was located between the desk and the bed on the window side of the room. A small flaming source was used to ignite the fuel package. Figure 2.3-6 shows the ignition fuel package. An electric match, as described in Table 2.3-5, was used to ignite the fuel package.

Figure 2.3-6. Photograph of ignition fuel package positioned between the desk and the bed closest to the window

Table 2.3-5. Ignition fuel package

Item	Materials	Dimensions			Mass
		W (m)	L (m)	H (m)	(kg)
Waste Basket	Wicker	0.25-0.20 (dia)		0.2	0.2
Newspaper	Newsprint	0.3	0.4		1.0
Notebook paper	Paper sheets, crumpled & straight	0.22	0.28		0.2
2- Plastic "sports" drink bottles, no lids .592 L (20 oz) size	PETE	0.08 (dia)		0.2	0.1
Electric Match	Paper matchbook with 20 matches, with ni-chrome wire coil	0.04	0.05	0.01	
Total					1.5

The total mass of fuel in each dorm room was approximately 614 kg (1350 lbs). Dividing the mass by the area of the room, 15.4 m^2 (166 ft^2), yields an averaged fuel load of nearly 40 kg/m^2 (8.1 lbs/ft^2).

2.4 Instrumentation

Five main instrument positions were used for each experiment. Two positions (1 and 2) were located in the room of fire origin as shown in Figure 2.4-1. The other three positions (3 through 5) were located in the corridor and remained the same for all five of the experiments, see Figure 2.4-2. Each position had a vertical array of 0.51 mm (0.02 in) nominal diameter bare bead, Type K thermocouples. In each array a thermocouple was located 25 mm, 0.305 m, 0.610 m, 0.910 m, 1.22 m, 1.52 m, 1.83 m, and 2.13m (1 in, 1 ft, 2 ft, 3ft, 4 ft, 5 ft, 6 ft, and 7 ft) below the ceiling.

Thermocouple arrays in the corridor were along the centerline of corridor. Thermocouples were also installed adjacent to the sprinkler and smoke alarm located in the dorm rooms.

Gas concentrations were sampled at two different positions. In each experiment, the sampling line inlets were located adjacent to TC Array 1 and TC Array 4 at 1.52 m (5.0 ft) above the floor. The sampling lines were connected to calibrated vacuum pumps which moved the gas samples through a conditioning system, which removed particulate and moisture from the gas sample.

The conditioning system consisted of two filtered cold traps inline to remove soot particles and moisture from the gas sample. The sampling line had a nominal internal diameter of 9.5 mm. The cold traps were approximately 190 mm long with an internal diameter of 60 mm for an approximate volume of 500 cm^3 for each trap. The sampling system volume was minimized to avoid damping out important peak values in the gas concentration measurements. The volume of each cold trap was reduced by approximately 70 % with the addition of glass wool and glass beads [3].

The dry gas samples were then drawn through a series of gas analyzers, which were installed in the room next to "Dorm Room 2". This room was sealed and protected from thermal and fire gas infiltration. In all of the experiments, oxygen was measured using paramagnetic analyzers and carbon monoxide and carbon dioxide were measured using non-dispersive infrared (NDIR) analyzers. The exhaust line from the analyzers discharged outside of the building.

The length of the sample lines and the conditioning system resulted in some delay in the measurements. The delay times were determined by discharging carbon dioxide near the entry to the sample line and measuring the time from discharge until the response from the oxygen and carbon dioxide instruments were observed. These delay times are accounted for in the data presented in this report.

13

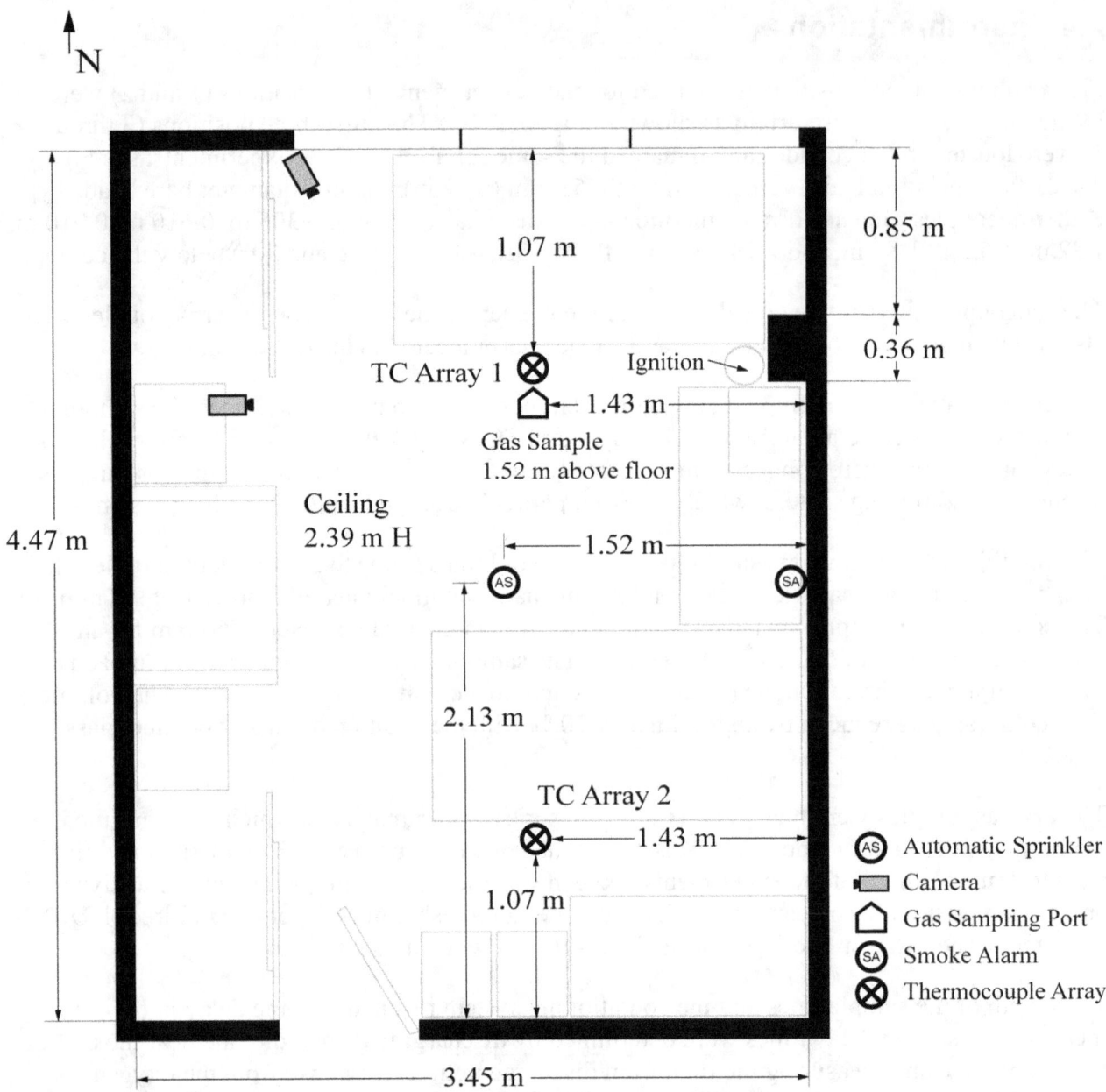

N

0.85 m

0.36 m

TC Array 1 ⊗

Ignition

1.07 m

⌂— 1.43 m

Gas Sample
1.52 m above floor

Ceiling
2.39 m H

4.47 m

1.52 m

(AS)   (SA)

2.13 m

TC Array 2

⊗—— 1.43 m——

1.07 m

(AS) Automatic Sprinkler
      Camera
⌂   Gas Sampling Port
(SA) Smoke Alarm
⊗   Thermocouple Array

3.45 m

Figure 2.4-1. Floor plan of dorm room with instrumentation locations

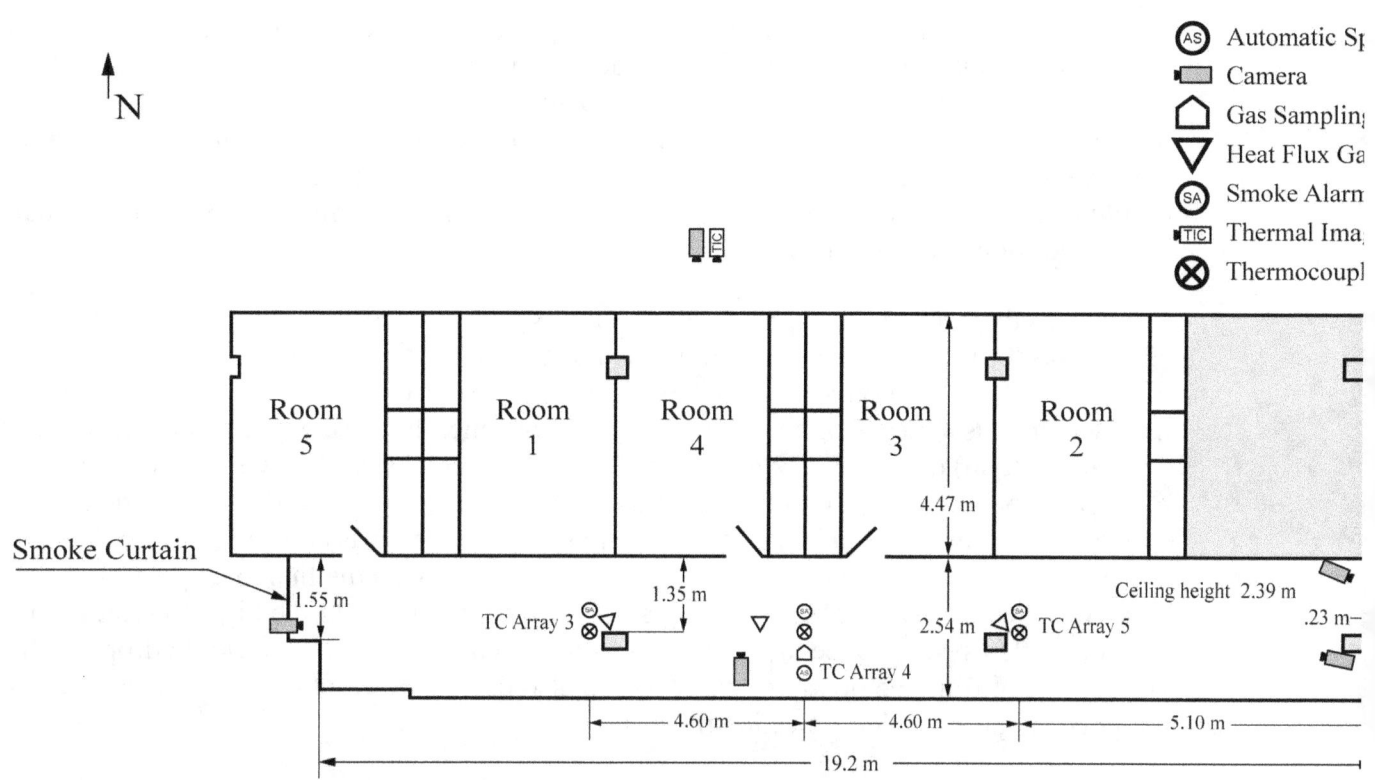

Figure 2.4-2. Floor plan of corridor with instrumentation locations

Three pairs of Schmidt Boelter total heat flux gauges were installed near the TC arrays positioned in the corridor. The heat flux gauges that were located next to TC Array 4, the center position of the corridor, had a design heat flux range up to 227 kW/m^2 (20 Btu/ft^2·s). The pair of heat flux gauges located next to TC Array 3, the west side of the corridor, had a design heat flux range up to 114 kW/m^2 (10 Btu/ft^2·s). The pair of heat flux gauges, installed next to TC array 5, the east side of the corridor, had a design heat flux range up to 57 kW/m^2 (5 Btu/ft^2·s).

Each pair of gauges consisted of one gauge facing the ceiling with another gauge facing the room of fire origin. The height of the gauges facing the ceiling was approximately 0.91 m (3 ft) above the corridor floor or 1.48 m (4.85 ft) below the ceiling. The height of the gauges positioned horizontally toward the fire were approximately 0.86 m (2.83 ft) above the corridor floor or 1.53 m (5 ft) below the ceiling.

Commercially available, battery powered, single station, ionization smoke alarms were used. A smoke alarm was mounted on the wall of each dorm room, as shown in Figure 2.4-1. The center of the alarm case was located 150 mm (6 in) below the ceiling. The alarms in the corridor were mounted under the ceiling at the locations shown in Figure 2.4-2. Each alarm was separately connected to the data acquisition system. The voltage change, as measured across the battery terminals at its alarm point, served as the data marker for the alarm time. New smoke alarms were used for each experiment.

Commercially available quick response sprinklers with a listed activation temperature of 74 °C (165 °F) were installed in each of the dorm rooms and in the center of the corridor adjacent to TC array 4. In the non-sprinklered experiments, (Experiments 1, 4, and 5), sprinklers were installed only as a means of obtaining an activation time. In these experiments, the nominal 12.7 mm (1/2 in) diameter, 152 mm (6 in) long drop nipple was filled with water and then the pipe was capped with an air pressure fitting. A 6 mm (0.25 in) outer diameter copper tube connected the sprinkler pipe to an air pressure sensor, which was connected to the data acquisition system. A valved tee fitting in the air line allowed the line to be pressurized to approximately 20 psi. When the line was pressurized, the valve to the high pressure source was closed. In this system, once the thermal element fused, the air pressure would drop and the voltage signal from the pressure sensor would also decrease. The time of this decrease was recorded as the sprinkler activation time.

In the sprinklered experiments, (Experiments 2 and 3), the sprinklers were connected to a water supply and allowed to operate normally. In the sprinklered experiments, the nominal 12.7 mm (½ in) pipe drops were connected via a nominal 1 inch pipe to a nominal 1 ½ single jacket hose line supplied by a fire engine. An in-line flow meter was located in the hose line. This flow meter was read manually during the experiments and the flow rates noted. Prior to installation in the building, the "sprinkler system" was calibrated for flow with an open sprinkler. This allowed the pump operator to pre-set the pump control to provide the desired flow rate, which was approximately 1.3 L/s (20 gpm).

2.5 Uncertainty Analysis

There are different components of uncertainty in the length, temperature, heat flux, gas concentrations, mass and flow rate provided in this report. Uncertainties are grouped into two categories according to the method used to estimate them. Type A uncertainties are those which are evaluated by statistical methods, and Type B are those which are evaluated by other means [4]. Type B analysis of systematic uncertainties involves estimating the upper (+ a) and lower (− a) limits for the quantity in question such that the probability that the value would be in the interval (± a) is very close to 100 %. For some of these components, such as the zero and calibration elements, uncertainties are derived from instrument specifications. Here uncertainty is reported as the expanded relative uncertainty with an expansion factor of 2 (i.e. 2σ).

Each length measurement was taken carefully. Length measurements such as the room dimensions, instrumentation array locations and furniture placement were made with steel tape measures with a resolution of ± 0.5 mm (0.02 in). However, conditions affecting the measurement, such as levelness or tautness of the device, yield an estimated uncertainty of ± 0.5 % for measurements in the 0.0 m (0 ft) to 3.0 m (9.8 ft) range. Some issues, such as "soft" edges on the upholstered furniture, or longer distances in excess of 3.0 m (9.8 ft) result in an estimated total expanded uncertainty of ± 1.0 %.

The standard uncertainty in temperature of the thermocouple wire itself is ± 2.2 °C at 277 °C and increases to ± 9.5 °C at 871 °C as determined by the wire manufacturer [5]. The variation of the temperature in the environment surrounding the thermocouple is known to be much greater than that of the wire uncertainty [6, 7]. Small diameter thermocouples were used to limit the impact of radiative heating and cooling. The estimated total expanded uncertainty for temperature in these experiments is ± 15 %.

In this study, total heat flux measurements were made with water-cooled Schimidt-Bolter gauges. The manufacturer reports a ± 3 % calibration expanded uncertainty for these devices [8]. Results from an international study on total heat flux gauge calibration and response demonstrated that the expanded uncertainty of a Schmidt-Boelter gauge is typically ± 8 % [9].

The gas measurement instruments and sampling system used in this series of experiments have demonstrated an expanded relative uncertainty of ± 1 % when compared with calibration span gas volume fractions [3, 10]. Given the limited set of sampling points in these experiments, an estimated uncertainty of ± 10 % is applied to the results.

Water flow rate was measured with a spring and piston type inline flow meter with a range from 0 L/s to 6.3 L/s (0 gpm to 100 gpm) The measuring accuracy per the manufacturer is 2.5 % full scale at mid range and 4.0 % of full scale over the entire scale range and repeatability of 1 % of full scale [11].

The load cell used to weigh the fuels prior to the experiments had a range of 0 kg (0 lbs) to 200 kg (440 lbs) with a resolution of a 0.05 kg (0.11 lb) and a calibration uncertainty within 1 % [12]. The expanded uncertainty is estimated to be ± 5 %.

17

In the following sections, the measurements are presented in graphic and tabular form. In the graphs, an error bar represents the estimated expanded uncertainty of the measurement.

2.6 Experimental Procedure

Prior to ignition in each experiment, a computerized data acquisition system was started to collect the temperature, gas concentration, heat flux and smoke alarm data. Data were collected from each instrument every 2 s. Video cameras recording the experiment were also started at this time.

After at least 60 s of background data were collected, a resistance heated book of matches was used to ignite the materials in the waste basket between the bed closest to the window and the adjacent desk. The fire growth was observed via monitors connected to the video cameras. After the experimental objectives were met, fire fighters entered the corridor from the door on the east end and suppressed the fire as needed.

Four different experimental conditions were used as shown in the table below. Three experiments did not have active sprinklers installed. Experiment 1 had the door between the dorm room and the corridor closed, while Experiment 4 and Experiment 5 had the door fully open. Experiment 2 and Experiment 3 used sprinklers with a water supply. Experiment 2 had the door between the dorm room and the corridor closed, while Experiment 3 had the door fully opened.

Table 2.6-1. Setup for the 5 experiments

Experiment	Door Position	Sprinkler Status
1	Closed	Non-Sprinklered
2	Closed	Sprinklered
3	Open	Sprinklered
4	Open	Non-Sprinklered
5	Open	Non-Sprinklered

3 Results

The results of the experiments include experiment timelines based on observations, smoke alarm and sprinkler activation times, temperature measurements, gas concentrations, heat flux measurements, photographs and videos.

3.1 Experiment 1: Closed Door, Non-Sprinklered

The objective of this experiment was to evaluate the impact that a closed door would have on the fire and the level of hazard developed in the room of origin and in the adjoining corridor.

3.1.1 Experiment Timeline

The timeline was developed from observations made during the experiment, review of the video of the experiment, and review of the data. Table 3.1-1 describes the level of fire development in the room. This can be compared with other measurements presented in following sections, such as changes in temperature, gas concentration or fire protection system response.

Table 3.1-1. Timeline for Experiment Number 1

Time (s)	Observations
0	Ignition
24	Smoke Alarm in fire room activated
35	Flames attached to bedding
40	Flames extended to pillow
45	Flames extended to desk chair
90	Smoke layer started to form
120	Tell-tale sprinkler activated
130	Blinds began to melt
150	Smoke layer approx. 1m below ceiling
160	Flames attached to desk
165	Plastic bookshelf melted and fell
205	Blinds fell
210	Flame size decreased
270	Smoke layer descended to floor
275	Flames no longer visible
345	Thermal plume no longer visible on IR Camera
555	Power off in room
610	Door open, Firefighters in
640	Fan in room ON

3.1.2 Smoke Alarm and Sprinkler Activation Times

The smoke alarm activation and sprinkler activation times are given in Table 3.1-2 and Table 3.1-3, respectively. The smoke alarm located in the dorm room activated at 24 s, while the fire was limited to the materials in the waste basket. The temperature at the thermocouple located near the smoke alarm indicated 52 °C (126 °F) at the time of activation. Even with the door between the dorm room and the corridor closed, enough smoke leaked into the corridor to activate all three of the smoke alarms over the course of 5 minutes following the activation of the alarm in the dorm room.

The fire generated enough heat to activate the sprinkler in the dorm room prior to the first smoke alarm in the corridor activating. The temperature measured adjacent to the sprinkler at the time of thermal activation was approximately 118 °C (244 °F). The sprinkler in this experiment did not have a water supply; it was pressurized with air. This time was documented to demonstrate when the sprinkler would have begun to have impact; however, this was a non-sprinklered experiment. The sprinkler installed in the center of the corridor, adjacent to TC array 4 did not activate.

Table 3.1-2. Dorm Room Experiment 1, Smoke Alarm Activation Times

Smoke Alarm	Location	Time (s)	Temperature (°C)
1	Dorm Room	24	52
2	West Corridor	160	27
3	Center Corridor	216	27
4	East Corridor	316	27

Table 3.1-3. Dorm Room Experiment 1, Sprinkler Activation Times

Sprinkler Activation	Location	Time (s)	Temperature (°C)
1	Dorm Room	120	118
2	Corridor	Did not activate	

3.1.3 Temperature Data

Temperatures from the 5 thermocouple arrays located in the dorm room (TC arrays 1 and 2) and in the corridor (TC arrays 3 through 5) are presented in Table 3.1-1 through Figure 3.1-5. Refer to Figure 2.4-1 for locations of the TC arrays.

Figure 3.1-1 shows the temperature data from the thermocouple array, in location 1, which was adjacent to the bed and closest to the source of ignition. The temperature 0.03 m (1 in) below the ceiling peaked at approximately 250 °C (480 °F), approximately 220 s after ignition. At this point in time, the flames had begun to decrease in size. After this point, all of the temperatures recorded at location 1 began to decrease. During this same period, the smoke layer extended to floor level and the flames were no longer visible in the room. At approximately 280 s after ignition, the temperatures at location 1 leveled off and remained constant for approximately 60 s. At 345 s, the thermal plume was no longer visible in the IR camera. After this point the temperatures continued to decrease. The small and abrupt temperature increase of the lower three thermocouples in the array at approximately 165 s after ignition was caused by the flames extending to the plastic crate book shelf on the desk, followed by the collapse of the book shelf and materials to the floor in the vicinity of the TC array 1.

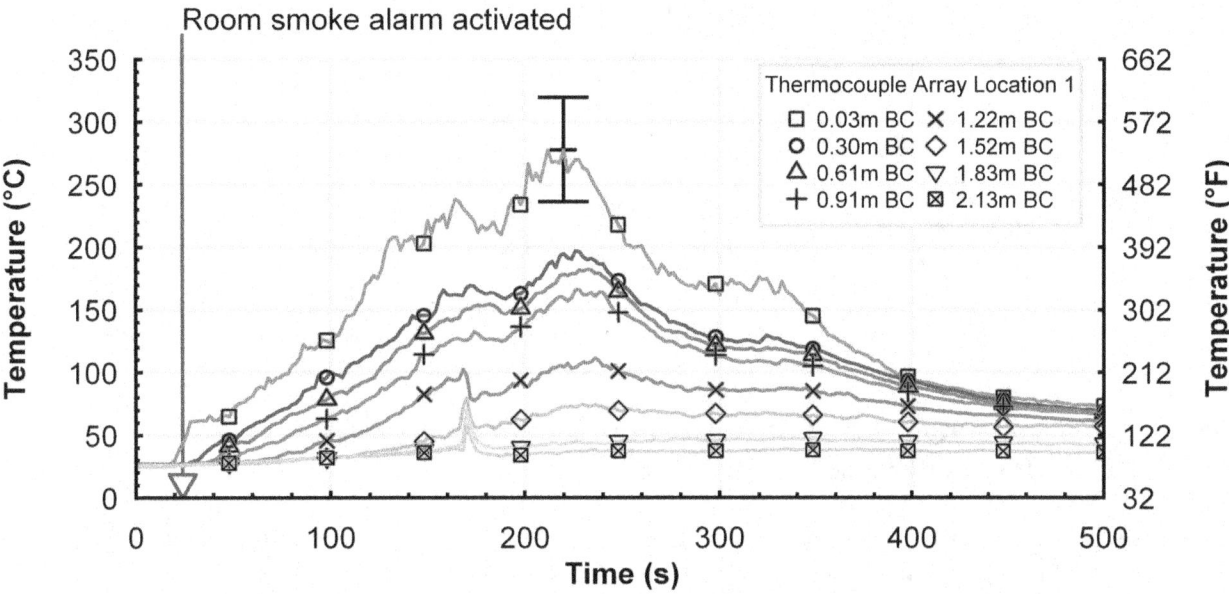

Figure 3.1-1. Temperature versus time data from thermocouple array 1 in Experiment 1, listed by distance below ceiling (BC)

As shown in Figure 3.1-2, the temperatures recorded at thermocouple array 2 had a very similar trend as those from thermocouple array 1. The temperature measurement at array 2 was approximately 50 °C (122 °F) less that the corresponding thermocouple in array 1, except for the thermocouple closest to the ceiling, which was somewhat lower in temperature.

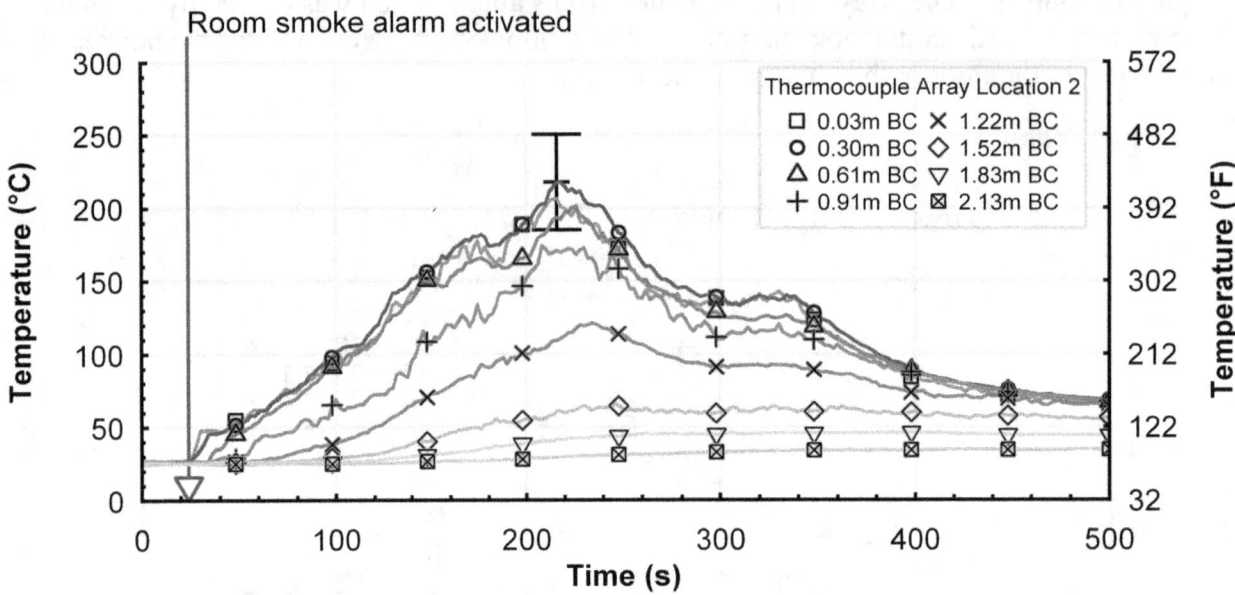

Figure 3.1-2. Temperature versus time data from thermocouple array 2 in Experiment 1, listed by distance below ceiling (BC)

The temperature time histories from the three vertical thermocouple arrays located along the center line of the corridor are given in Figure 3.1-3 through Figure 3.1-5. The closed door to the dorm room limited the flow of hot gases into the corridor. As a result, the temperatures in the corridor did not change significantly during the course of the experiment.

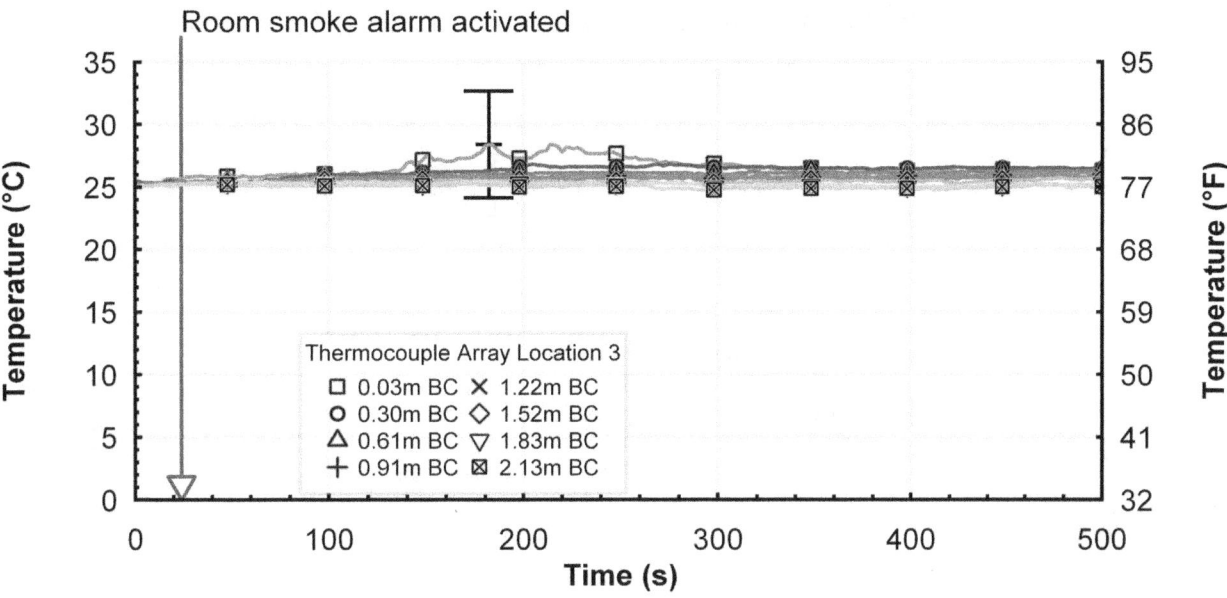

Figure 3.1-3. Temperature versus time data from thermocouple array 3 in Experiment 1, listed by distance below ceiling (BC)

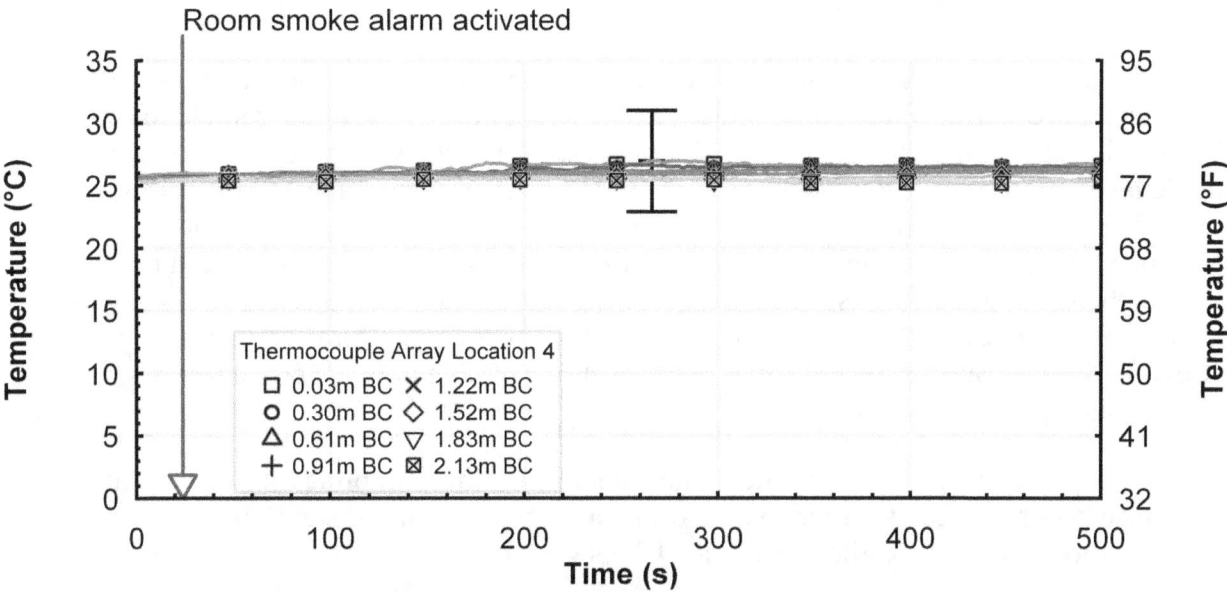

Figure 3.1-4. Temperature versus time data from thermocouple array 4 in Experiment 1, listed by distance below ceiling (BC)

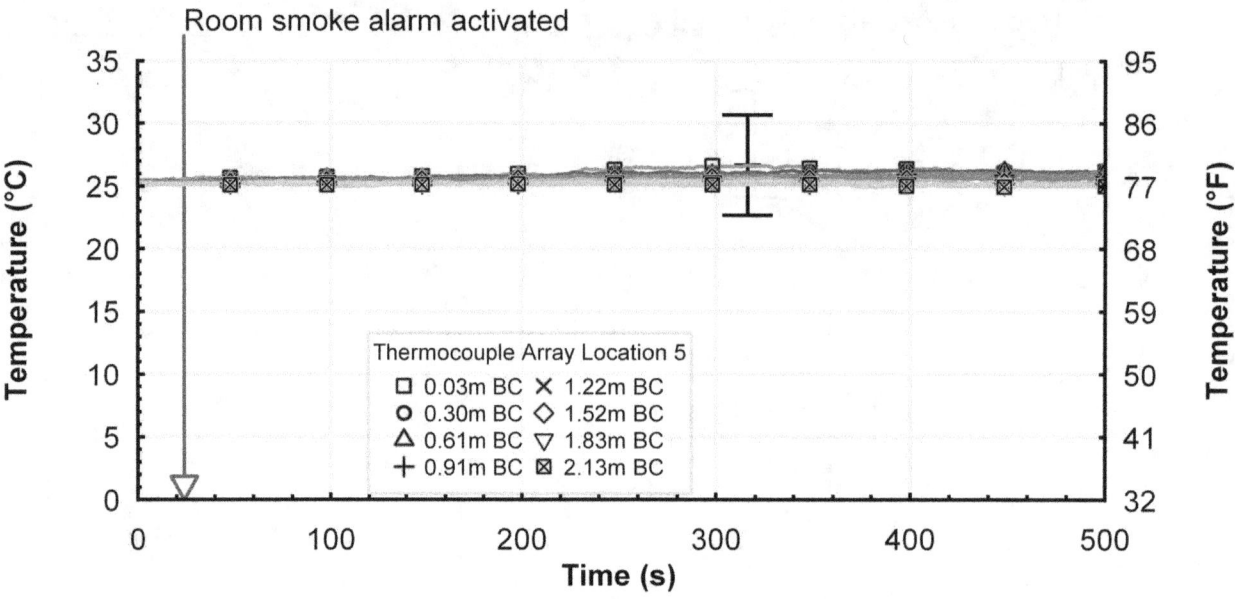

Figure 3.1-5. Temperature versus time data from thermocouple array 5 in Experiment 1, listed by distance below ceiling (BC)

3.1.4 Gas Concentrations

The gas concentrations were sampled in two locations, one in the dorm room adjacent to TC array 1 and one centered in the corridor adjacent to TC array 4. Both of the sample locations were positioned 1.52 m (5.0 ft) above the floor. Figure 3.1-6 shows the oxygen and carbon dioxide levels on the upper graph with a range of 0 % to 25 % by volume and the carbon monoxide is shown on the lower graph which has a range of 0 % to 1 % by volume. At approximately 350 s, the oxygen concentration dropped below 15 %. It was around this same time that the thermal plume was no longer visible in IR camera view. Based on the temperature data presented above and the information from the IR camera, it is estimated that the flaming fire self-extinguished at approximately 350 s. The level of carbon monoxide continued to increase after the flaming combustion appeared to cease.

Figure 3.1-7 shows the gas concentration data from the corridor sampling location. Again, due to the closed door between the room of fire origin and the corridor, no significant changes in the oxygen, carbon dioxide, or carbon monoxide levels were measured.

24

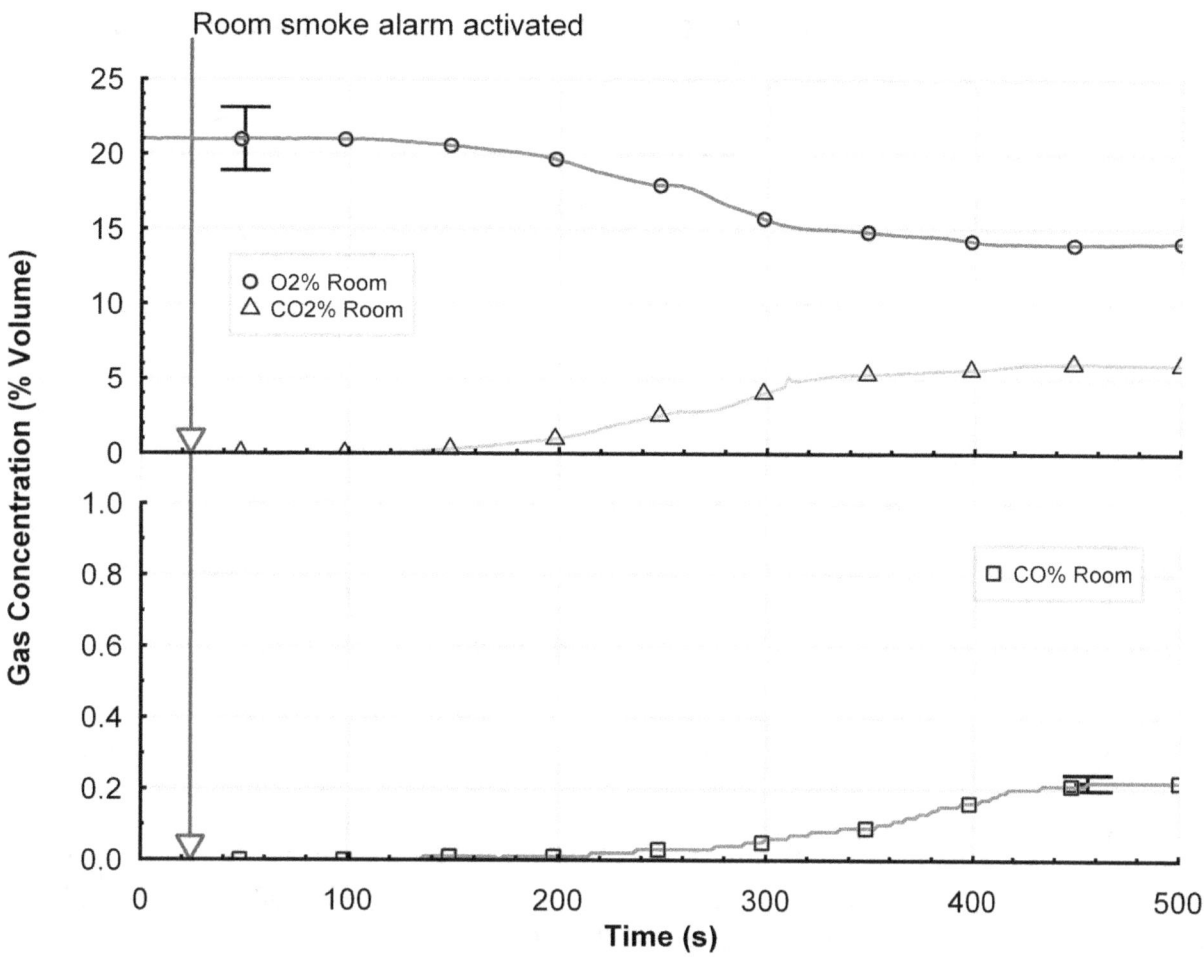

Figure 3.1-6. Relative gas concentrations versus time for the gas sampling point at 1.52 m above the floor at location 1 in Experiment 1

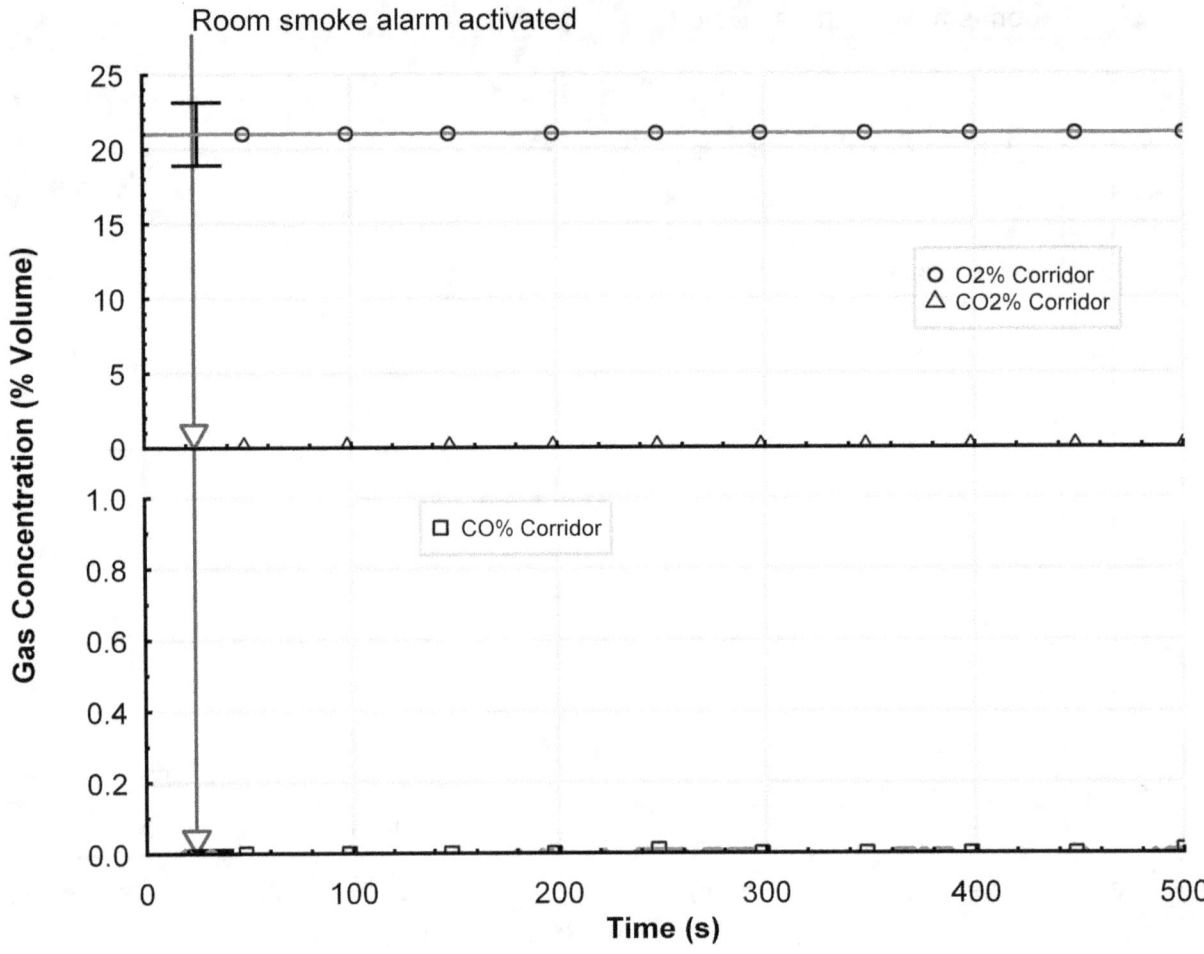

Figure 3.1-7. Relative gas concentrations versus time for the gas sampling at 1.52 m above the floor at location 4 in Experiment 1

3.1.5 Heat Flux Data

The heat flux data from the three sets of gauges in the corridor is shown in Figure 3.1-8. Adjacent to each thermocouple array in the corridor there were two total heat flux gauges, one aimed at the ceiling (Vertical) and one aimed toward the fire room (Horizontal). The heat flux from the pair of gauges located near array 4, as well as those closest to the room of origin, increased to approximately 0.5 kW/m^2. As a point of reference, the heat flux from the sun on a sunny day at ground level is approximately 1 kW/m^2.

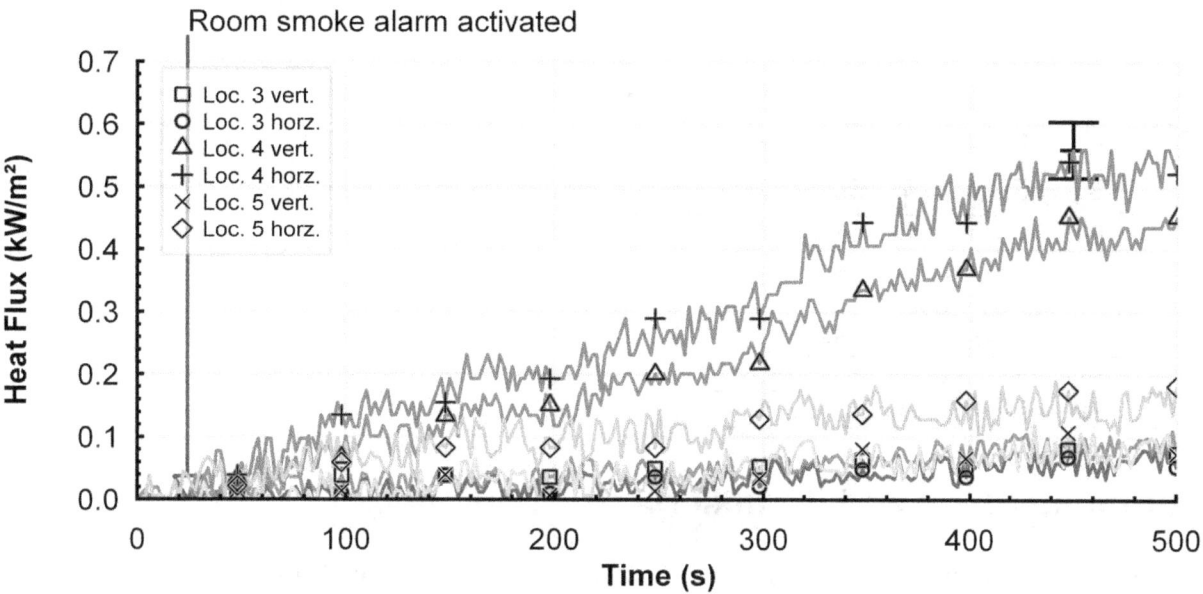

Figure 3.1-8. Heat flux versus time data for the heat flux gauges at locations 3, 4, and 5 in the orientations up (vert) and sideways (horz)

3.2 Experiment 2: Closed Door, Sprinklered

The objective of this experiment was to examine the impact of an automatic sprinkler on a fire in a dorm room with a closed door. This experiment enables direct comparison between the results from Experiment 1 and Experiment 2.

3.2.1 Experiment Timeline

The timeline was developed from observations made during the experiment, review of the video of the experiment, and review of the data. Table 3.2-1 provides a reference to the level of fire development in the room. This can be compared with other measurements presented in following sections, such as changes in temperature, gas concentration or fire protection system response.

Table 3.2-1. Timeline for Experiment Number 2

Time (s)	Observations
0	Ignition
12	Smoke Alarm in fire room activated
16	Flames attached to bedding
31	Flames extended to pillow
60	Smoke layer began to form
90	Flames extended to desk
105	Sprinkler activated
125	Flames extinguished

3.2.2 Smoke Alarm and Sprinkler Activation Times

The smoke alarm activation and sprinkler activation times are given in Table 3.2-2 and Table 3.2-3, respectively. The smoke alarm located in the dorm room activated at 12 s, while the fire was limited to the materials in the waste basket. The temperature at the thermocouple located near the smoke alarm indicated 32 °C (90 °F) at the time of activation. Due to the door being closed in this test and the rapid reduction of the fire hazard by the sprinkler, not enough smoke leaked into the corridor to activate the three remaining smoke alarms.

The fire generated enough heat to activate the sprinkler in the dorm room at 105 s after ignition. At this point the temperature adjacent to the sprinkler was approximately 119 °C (246 °F). In this experiment, the sprinkler did have a water supply and was flowing water at approximately 1.3 L/s (20 gpm). Within 30 s after activation of the sprinkler, the fire was almost completely extinguished as indicated by the video and temperature records.

The sprinkler located in the corridor was not exposed to enough thermal energy to activate. This was due to the door between the dorm room and the corridor being closed, just as in Experiment 1.

Table 3.2-2. Dorm Room Experiment 2, Smoke Alarm Activation Times

Smoke Alarm	Location	Time (s)	Temperature (°C)
1	Dorm Room	12	32
2	West Corridor	Did not activate	
3	Center Corridor	Did not activate	
4	East Corridor	Did not activate	

Table 3.2-3. Dorm Room Experiment 2, Sprinkler Activation Times

Sprinkler Activation	Location	Time (s)	Temperature (°C)
1	Dorm Room	105	119
2	Corridor	Did not activate	

3.2.3 Temperature Data

Temperatures from the 5 thermocouple arrays located in the dorm room (TC arrays 1 and 2) and in the corridor (TC arrays 3 through 5) are presented in Figure 3.2-1 through Figure 3.2-5.

Figure 3.2-1 shows the temperature data from the thermocouple array in location 1, which was adjacent to the bed and closest to the source of ignition. The temperature 0.03 m (1 in) below the ceiling peaked at approximately 175 °C (347 °F), approximately 108 s after ignition. At this point in time, the sprinkler in the dorm room had activated and water began to suppress the fire. After this point, all of the temperatures recorded at location 1 rapidly decreased over a span of about 20 s. During this same period, the thermal plume diminished until it was no longer visible with the IR camera. Within 60 s of sprinkler activation, the temperatures throughout the dorm room equalized at approximately 30 °C (86 °F).

Figure 3.2-2 shows the temperature data for the thermocouple array in location 2, which was adjacent to the other bed and closest to the corridor door. This TC array was also in the fire room, thus the trends were consistent with that of TC array 1, excluding the maximum temperature. At approximately 108 s after ignition, the maximum temperature at thermocouple array 2 was 118 °C (244 °F).

The temperature time histories from the three vertical thermocouple arrays located along the center line of the corridor are given in Figure 3.2-3 through Figure 3.2-5. The closed door to the dorm room limited the flow of hot gases into the corridor. As a result, the temperatures in the corridor do not change significantly during the course of the experiment.

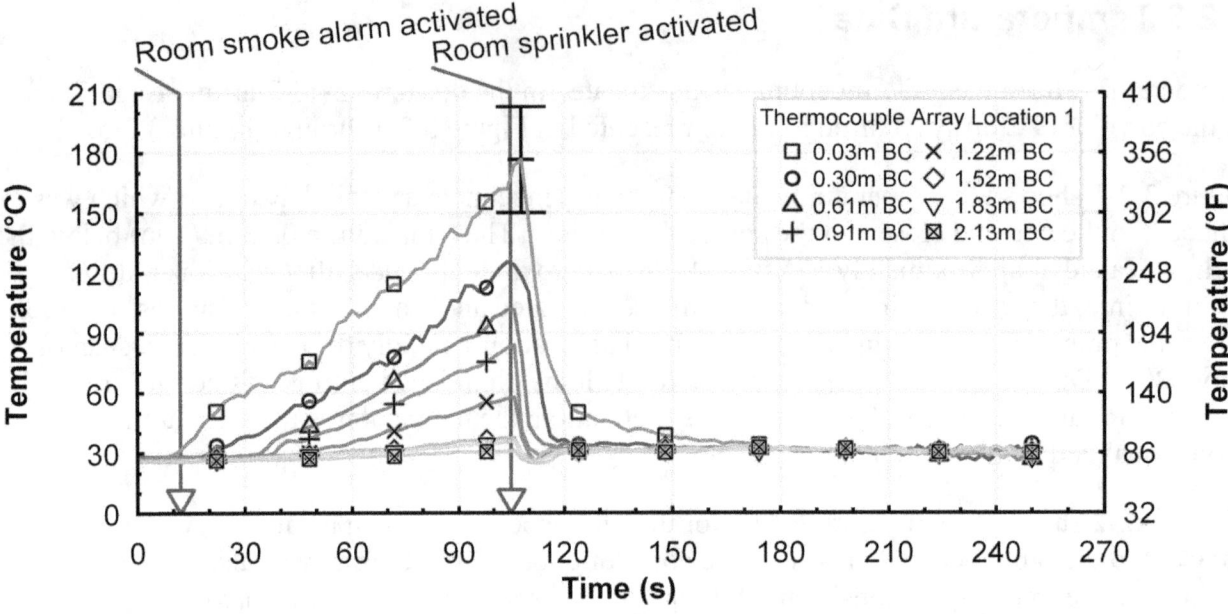

Figure 3.2-1. Temperature versus time data from thermocouple array 1 in Experiment 2, listed by distance below ceiling (BC)

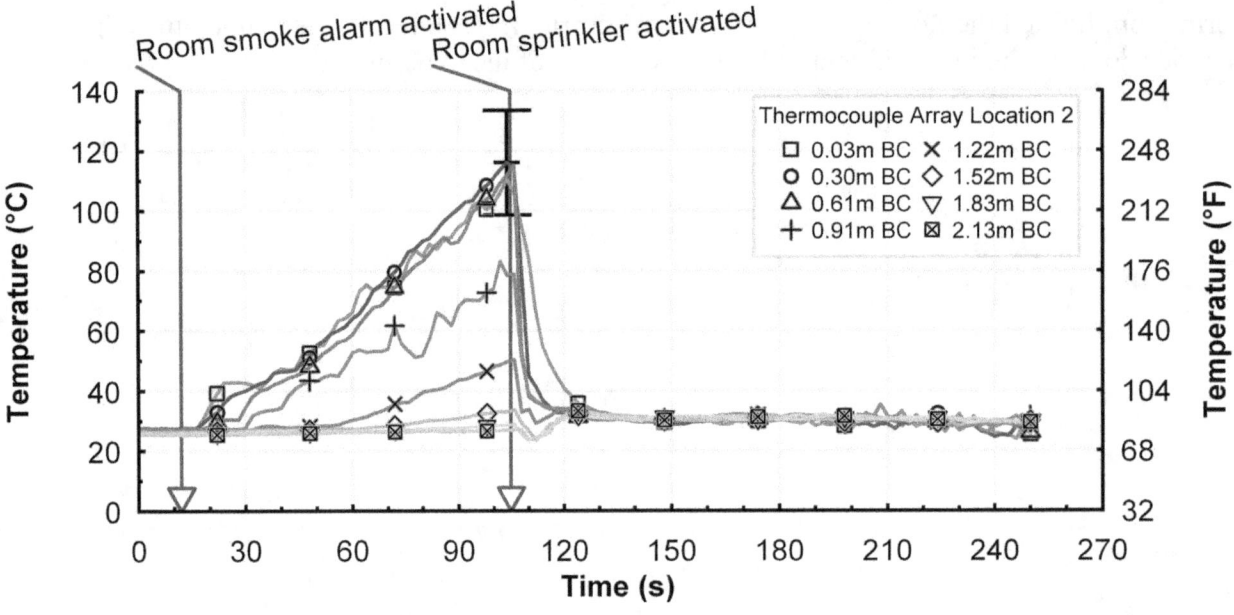

Figure 3.2-2. Temperature versus time data from thermocouple array 2 in Experiment 2, listed by distance below ceiling (BC)

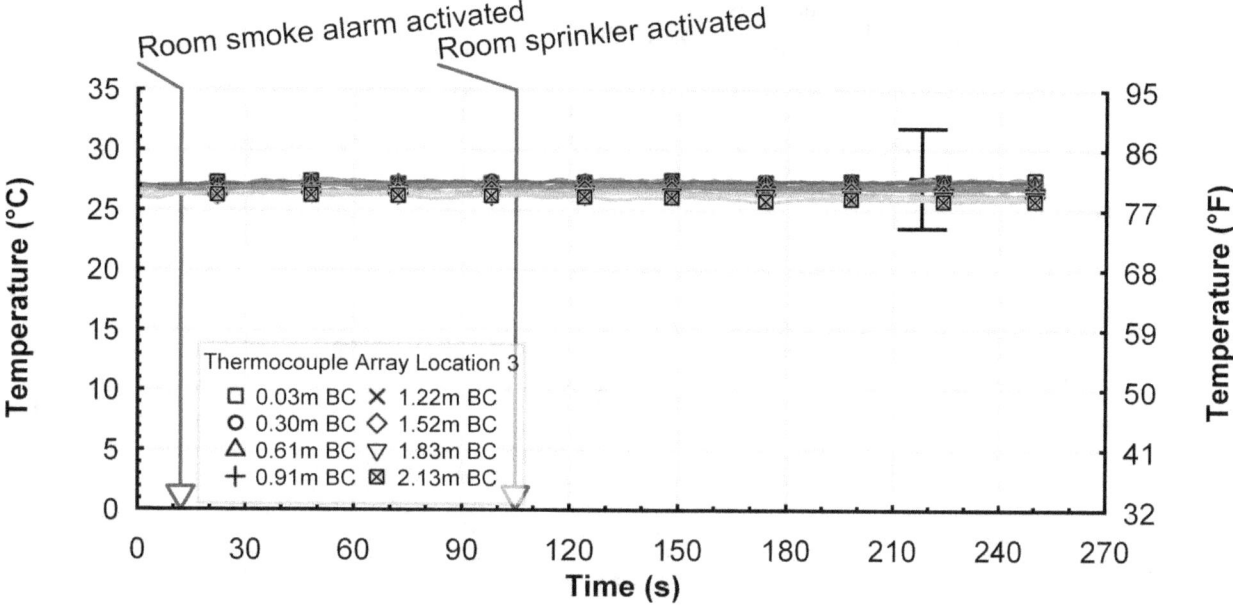

Figure 3.2-3. Temperature versus time data from thermocouple array 3 in Experiment 2, listed by distance below ceiling (BC)

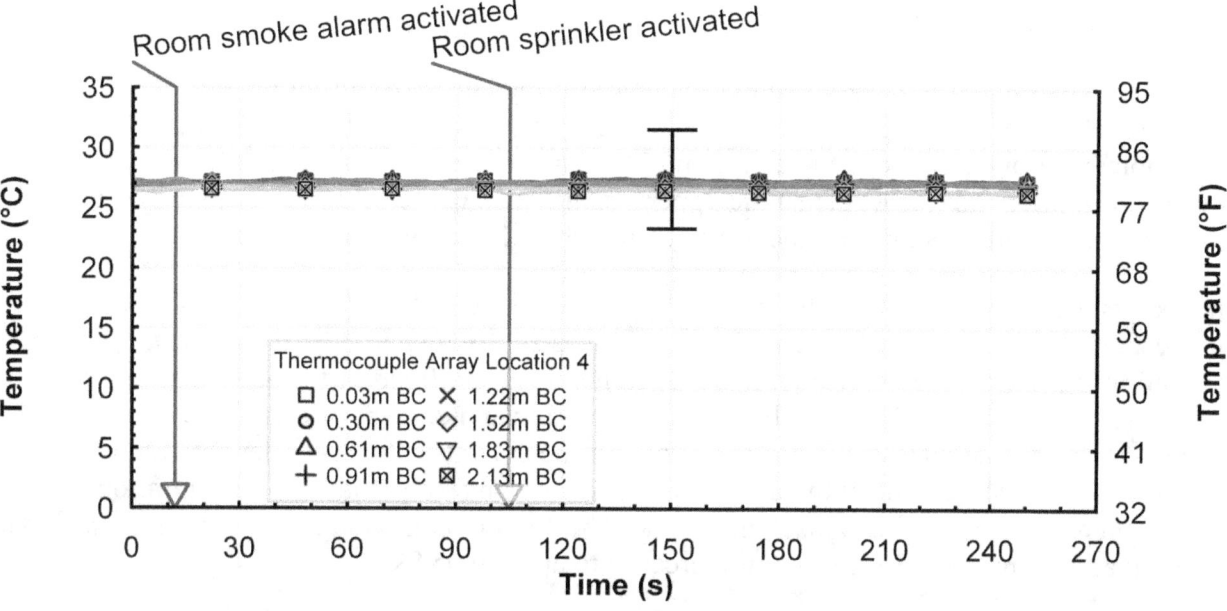

Figure 3.2-4. Temperature versus time data from thermocouple array 4 in Experiment 2, listed by distance below ceiling (BC)

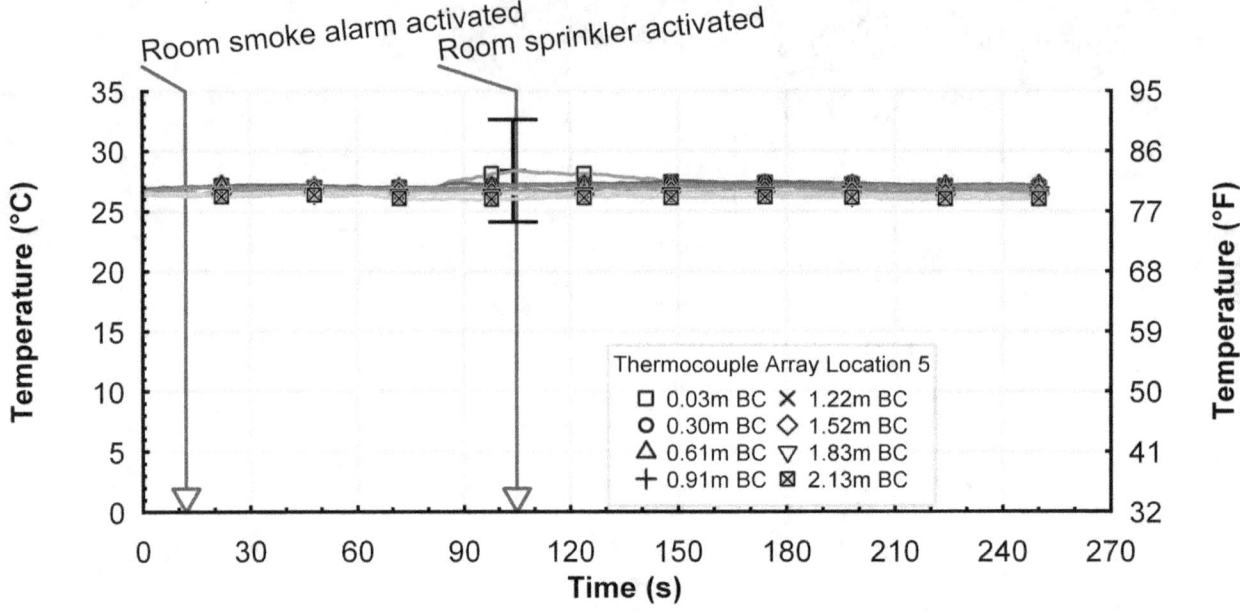

Figure 3.2-5. Temperature versus time data from thermocouple array 5 in Experiment 2, listed by distance below ceiling (BC)

3.2.4 Gas Concentrations

The gas concentrations were sampled in two locations, one in the dorm room adjacent to thermocouple array 1 and one centered in the corridor adjacent to thermocouple array 4. Both of the sample locations were positioned 1.52 m (5.0 ft) above the floor. Figure 3.2-6 shows the oxygen and carbon dioxide levels on the upper graph with a range of 0 % to 25 % by volume and the carbon monoxide is shown on the lower graph which has a range of 0 % to 1 % by volume.

The oxygen concentration in the dorm room remained above 19 % throughout the experiment. The decrease began at approximately 110 s. This was approximately the same time that the thermal plume was no longer visible in the IR camera and shortly after the automatic sprinkler in the dorm room activated. This may be due to the mixing and descent of combustion products which were above the inlet level of the gas sample line at the time of sprinkler activation. The carbon dioxide concentration at this location was related inversely with the concentration of oxygen, increasing from 0 % to just below 2 %. The carbon monoxide concentration remained at 0 % until approximately 180 s, where it increased to about 0.05 %.

The gas concentrations measured from the center of the corridor were consistent with each other in that they did not change. There was no significant change in any of the three gas concentrations because the fire hazard did not impact the conditions in the corridor due to the door to the dorm room being closed.

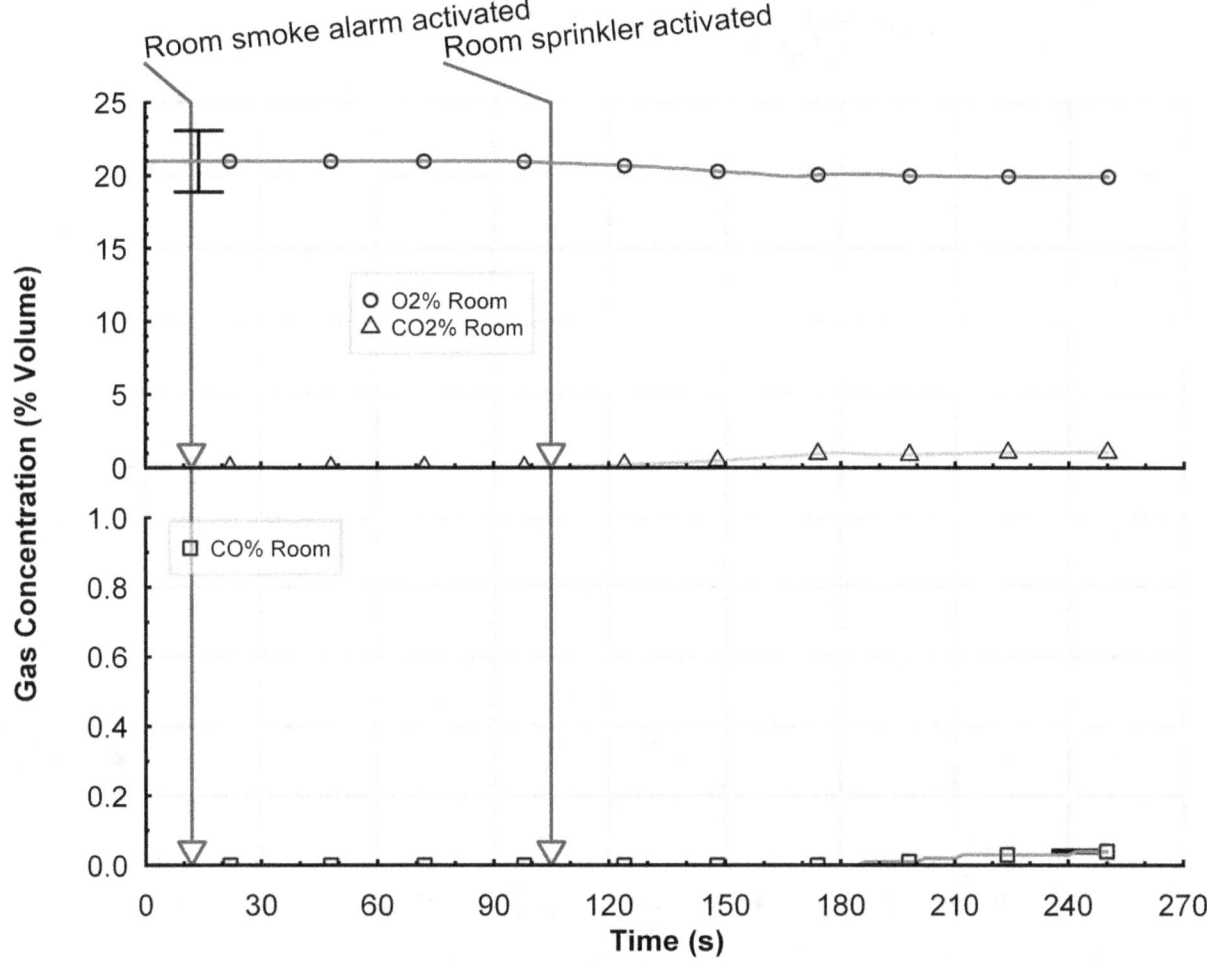

Figure 3.2-6. Relative gas concentrations versus time for the gas sampling at 1.52 m above the floor at location 1 in Experiment 2

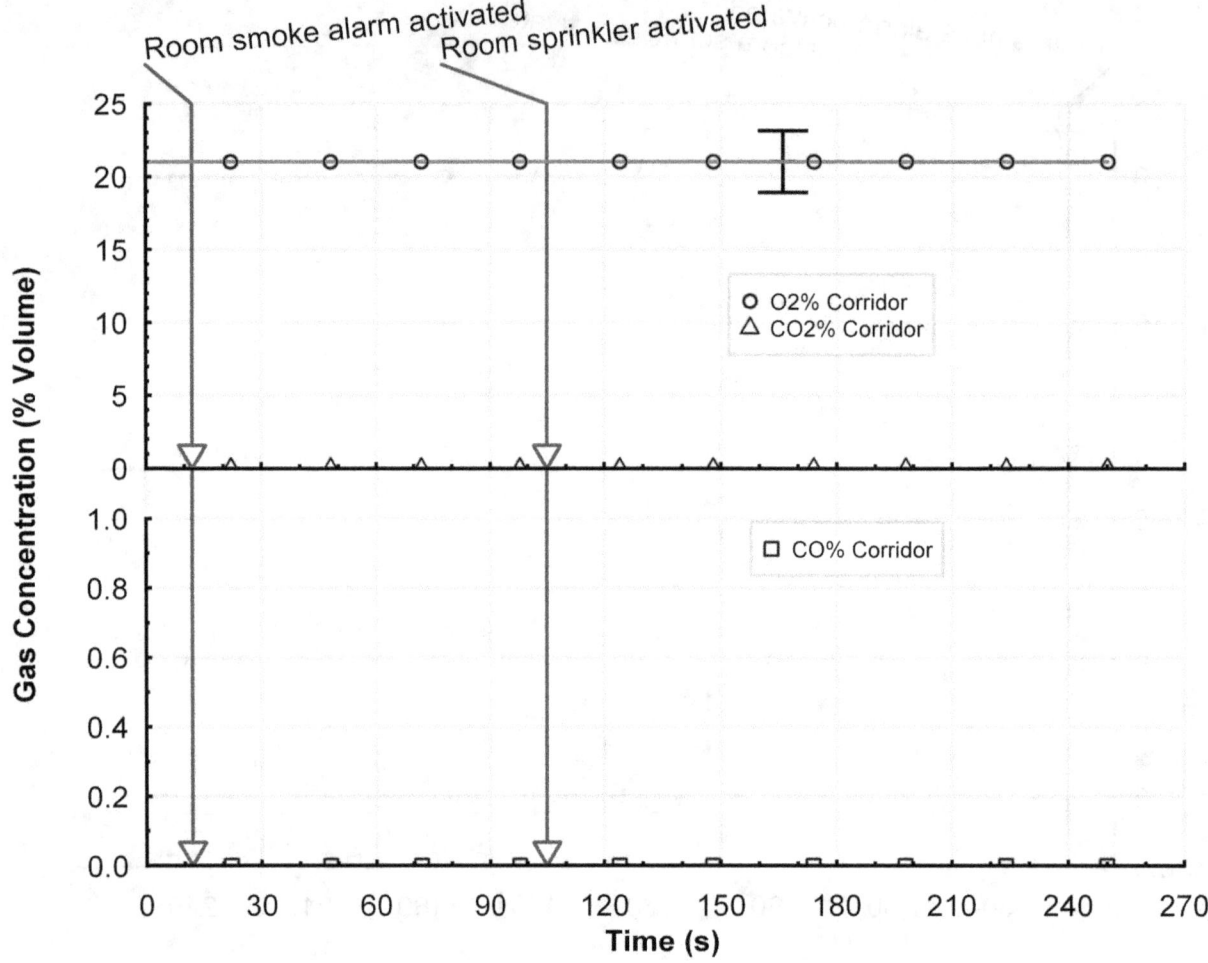

Figure 3.2-7. Relative gas concentrations versus time for the gas sampling at 1.52 m above the floor at location 4 in Experiment 2

3.2.5 Heat Flux Data

The heat flux data from the three sets of gauges in the corridor is shown in Figure 3.2-8. Overall, the heat flux throughout the corridor was similar and did not drift far from 0.0 kW/m^2.

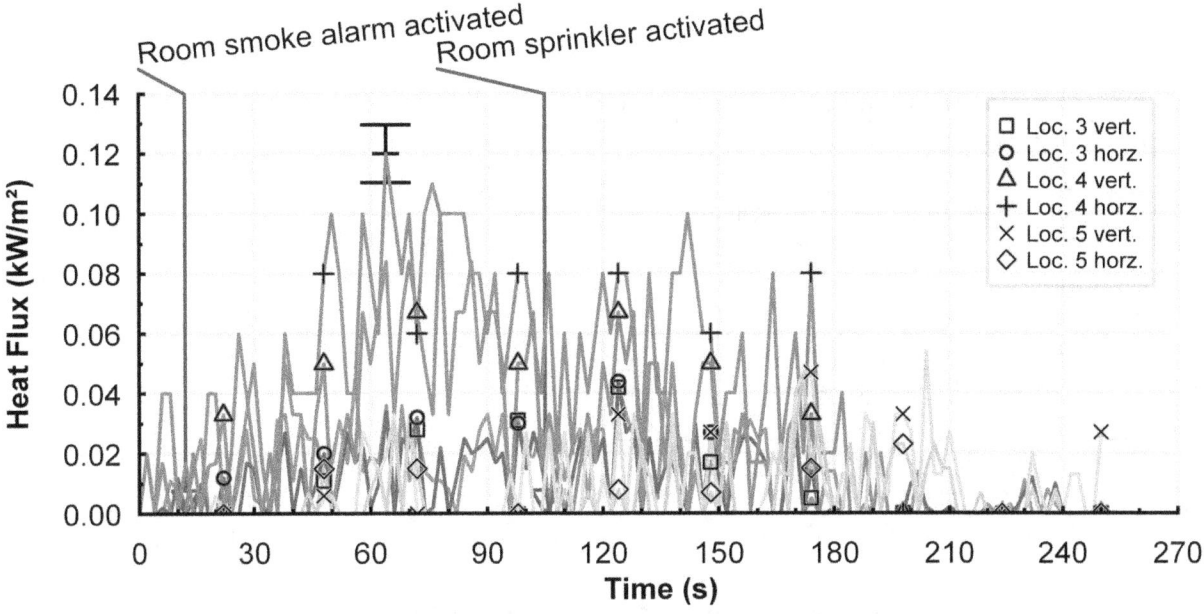

Figure 3.2-8. Heat flux versus time data for the heat flux gauges at locations 3, 4, and 5 in the orientations up (U) and sideways (S) in Experiment 2

3.3 Experiment 3: Open Door, Sprinklered

The objective of this experiment was to examine the impact of the automatic sprinkler alone on a fire in the dorm room without the benefit of the closed door between the dorm room and the corridor. The effect of the closed door is examined later in this report by direct comparison between Experiment 2 and 3.

3.3.1 Experiment Timeline

The timeline was developed from observations made during the experiment, review of the video of the experiment, and review of the data. Table 3.3-1 describes the level of fire development in the room. This can be compared with other measurements presented in following sections, such as changes in temperature, gas concentration or fire protection system response.

35

Table 3.3-1. Timeline for Experiment Number 3

Time (s)	Observations
0	Ignition
22	Smoke alarm in fire room activated
35	Flames attached to bedding
60	Fire extended to desk chair
65	Fire extended to pillow
70	Visible smoke layer started to form
85	Flames extended to desk
112	Sprinkler activation
115	Smoke visible in corridor
120	Fire extinguished

3.3.2 Smoke Alarm and Sprinkler Activation Times

The smoke alarm activation and sprinkler activation times are given in Table 3.3-2 and Table 3.3-3 respectively. The smoke alarm located in the dorm room activated at 22 s, while the fire was limited to the materials in the waste basket. The temperature at the thermocouple located near the smoke alarm was 46 °C (115 °F) at the time of activation. Due to the open door between the dorm room and the corridor, within 68 s after ignition, the three smoke alarms installed on the ceiling of the corridor activated.

The fire generated enough heat to activate the sprinkler in the dorm room at 112 s after ignition. At this point, the temperature adjacent to the sprinkler was approximately 112 °C (233 °F). In this experiment, the sprinkler had a water supply, flowing at approximately 1.3 L/s (20 gpm). Within 20 s after the activation of the sprinkler, the fire was almost completely extinguished.

In this experiment, even with the door open, the sprinkler located in the corridor was not exposed to enough thermal energy to activate. This was due to the rapid reduction of heat release rate from the fire and cooling of the combustion products caused by the activation of the single sprinkler in the dorm room.

Table 3.3-2. Dorm Room Experiment 3, Smoke Alarm Activation Times

Smoke Alarm	Location	Time (s)	Temperature (°C)
1	Dorm Room	22	46
2	West Corridor	68	27
3	Center Corridor	36	28
4	East Corridor	62	29

Table 3.3-3. Dorm Room Experiment 3, Sprinkler Activation Times

Sprinkler Activation	Location	Time (s)	Temperature (°C)
1	Dorm Room	112	112
2	Corridor	Did not activate	

3.3.3 Temperature Data

Temperatures from the 5 thermocouple arrays located in the dorm room (TC arrays 1 and 2) and in the corridor (TC arrays 3 through 5) are presented in Figure 3.3-1 through Figure 3.3-5.

Figure 3.3-1 shows the temperature data from the thermocouple array, in location 1, which was adjacent to the bed and closest to the source of ignition. The temperature 0.03 m (1 in) below the ceiling peaked at approximately 170 °C (340 °F), approximately 115 s after ignition, just a few seconds after the sprinkler activated. Within 20 s of sprinkler activation, the temperatures measured at TC array 1 had all equalized to approximately 30 °C (86 °F).

Figure 3.3-2 shows the temperature data for the thermocouple array in location 2, which was adjacent to the other bed and closest to the corridor door. Again, the trends from TC array 2 were consistent with that of TC array 1, with the exception of the temperature measured 0.03 m below the ceiling. At approximately 115 s after ignition, the maximum temperature 0.03 m below the ceiling at TC array 2 was just over 100 °C (212 °F), about 70 °C (160 °F) less that the same position in TC array 1 at this time. This was the only time and location that the temperatures between TC array 1 and 2 were significantly different. This difference was most likely due to the close proximity of TC array 1 to the thermal plume impingement area on the ceiling.

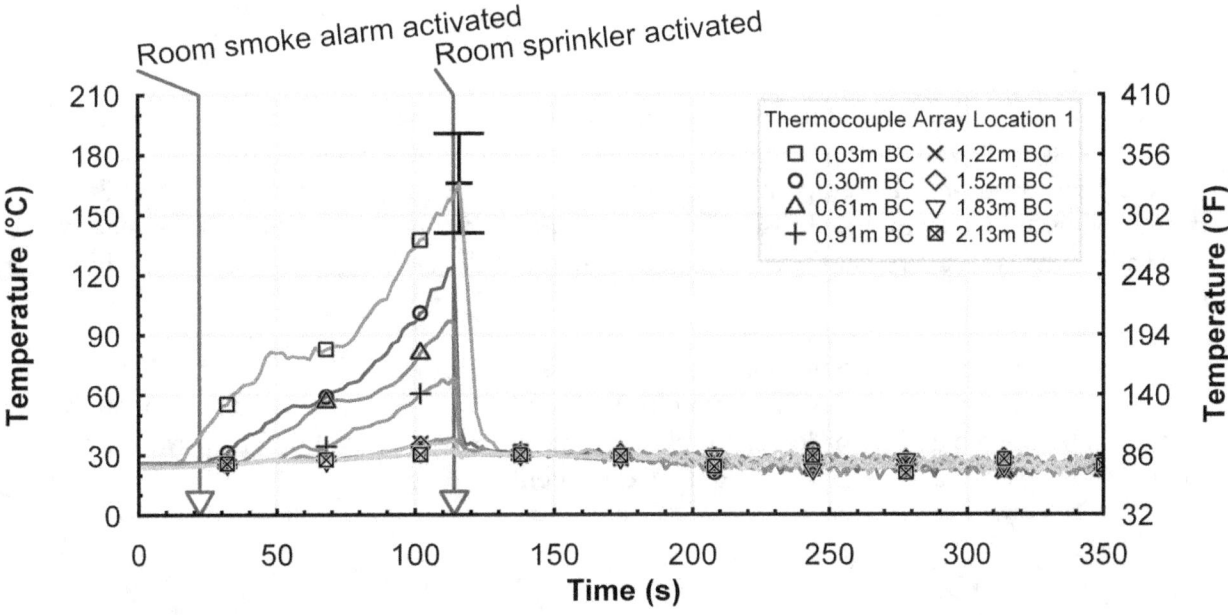

Figure 3.3-1. Temperature versus time data from thermocouple array 1 in Experiment 3, listed by distance below ceiling (BC)

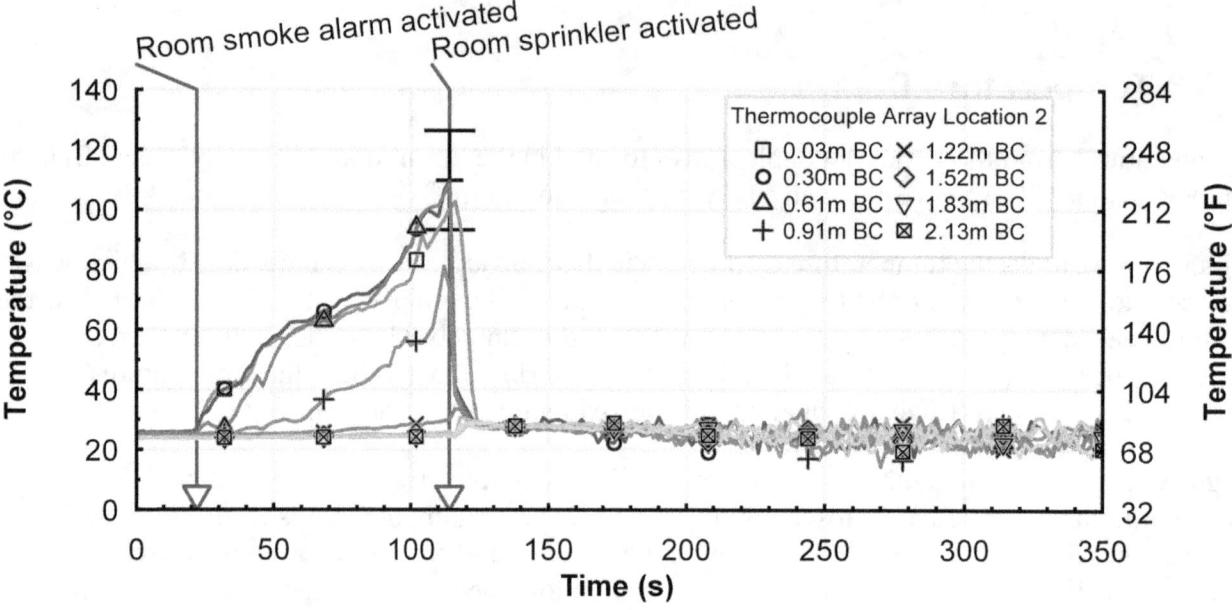

Figure 3.3-2. Temperature versus time data from thermocouple array 2 in Experiment 3, listed by distance below ceiling (BC)

The temperature time histories from the three vertical thermocouple arrays located along the center line of the corridor are given in Figure 3.3-3 through Figure 3.3-5. Unlike the previous two experiments, the open door to the dorm room allowed hot combustion products to flow into the corridor.

TC array 4 was the closest of the three arrays in the corridor to the open doorway of the burning room. As a result, the hot gases reached TC array 4 first and provided the highest temperature increase at that location. The temperature 0.03 m below the ceiling began to increase at 30 s after ignition as shown in Figure 3.3-4. The peak temperature in the corridor, approximately 70 °C (160 °F), was measured at the same location at about the same time that the sprinkler activated.

The only significant temperature increases at TC arrays 3 and 5 were limited to less than 20 °C (68 °F) above ambient at positions within 0.30 m (1 ft) of the ceiling. The temperatures 0.61 m (2 ft) below the ceiling or lower for all three of the arrays located in the corridor never exceeded 30 °C (86 °F) at any time during the experiment.

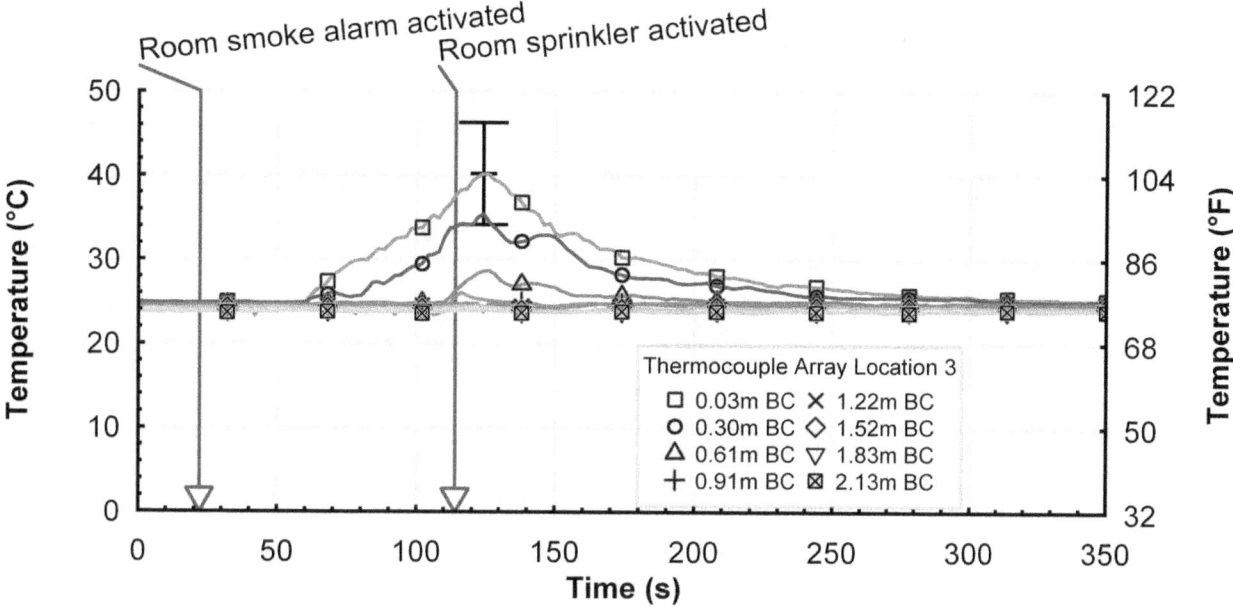

Figure 3.3-3. Temperature versus time data from thermocouple array 3 in Experiment 3, listed by distance below ceiling (BC)

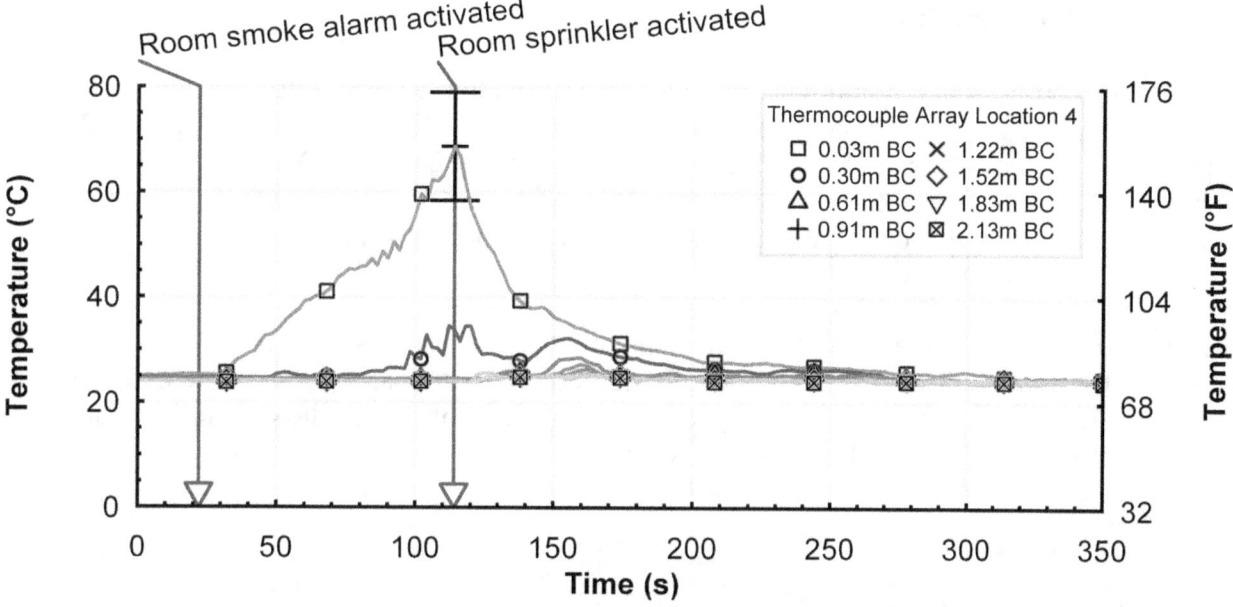

Figure 3.3-4. Temperature versus time data from thermocouple array 4 in Experiment 3, listed by distance below ceiling (BC)

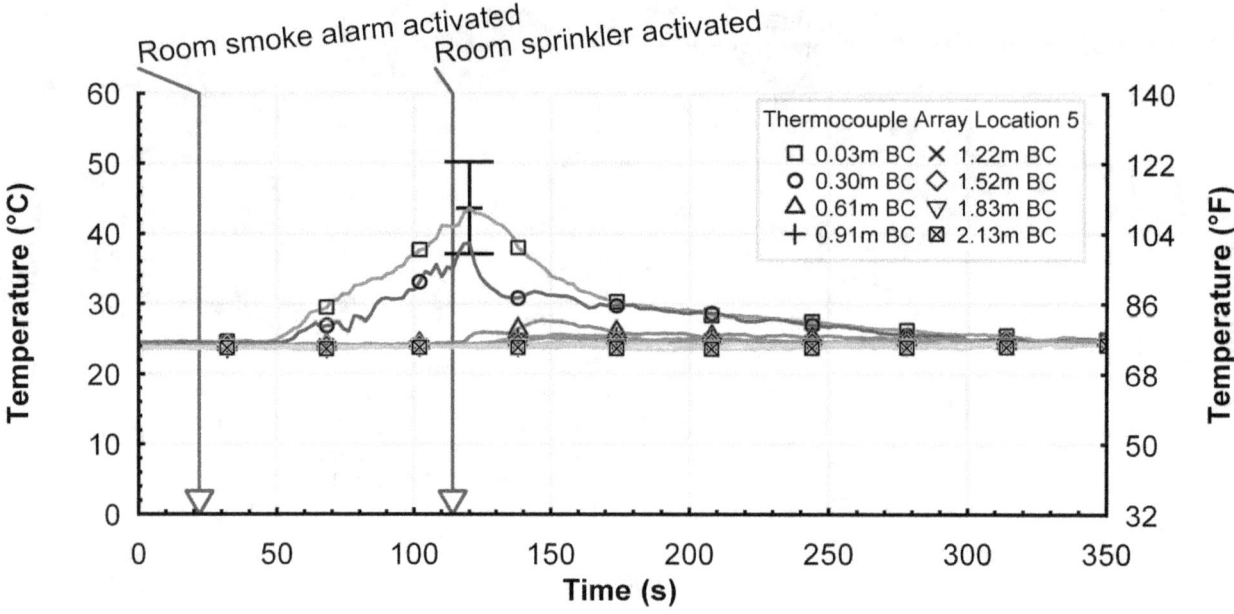

Figure 3.3-5. Temperature versus time data from thermocouple array 5 in Experiment 3, listed by distance below ceiling (BC)

3.3.4 Gas Concentrations

The gas concentrations were sampled in two locations, one in the dorm room adjacent to TC array 1 and one centered in the corridor adjacent to TC array 4, Figure 3.3-6 and Figure 3.3-7 respectively. Both of the sample locations were positioned 1.52 m (5.0 ft) above the floor. In each figure, the oxygen and carbon dioxide levels are given on the upper graph with a range of 0 % to 25 % by volume and the carbon monoxide measurement is shown on the lower graph which has a range of 0 % to 1 % by volume.

The oxygen concentration at both the dorm room and the corridor positions never measured less than 20 % by volume. The peak measures of carbon dioxide and carbon monoxide in the dorm room were 1.0 % and less than 0.1 % respectively. Increases of carbon dioxide and carbon monoxide at the corridor sampling position were negligible.

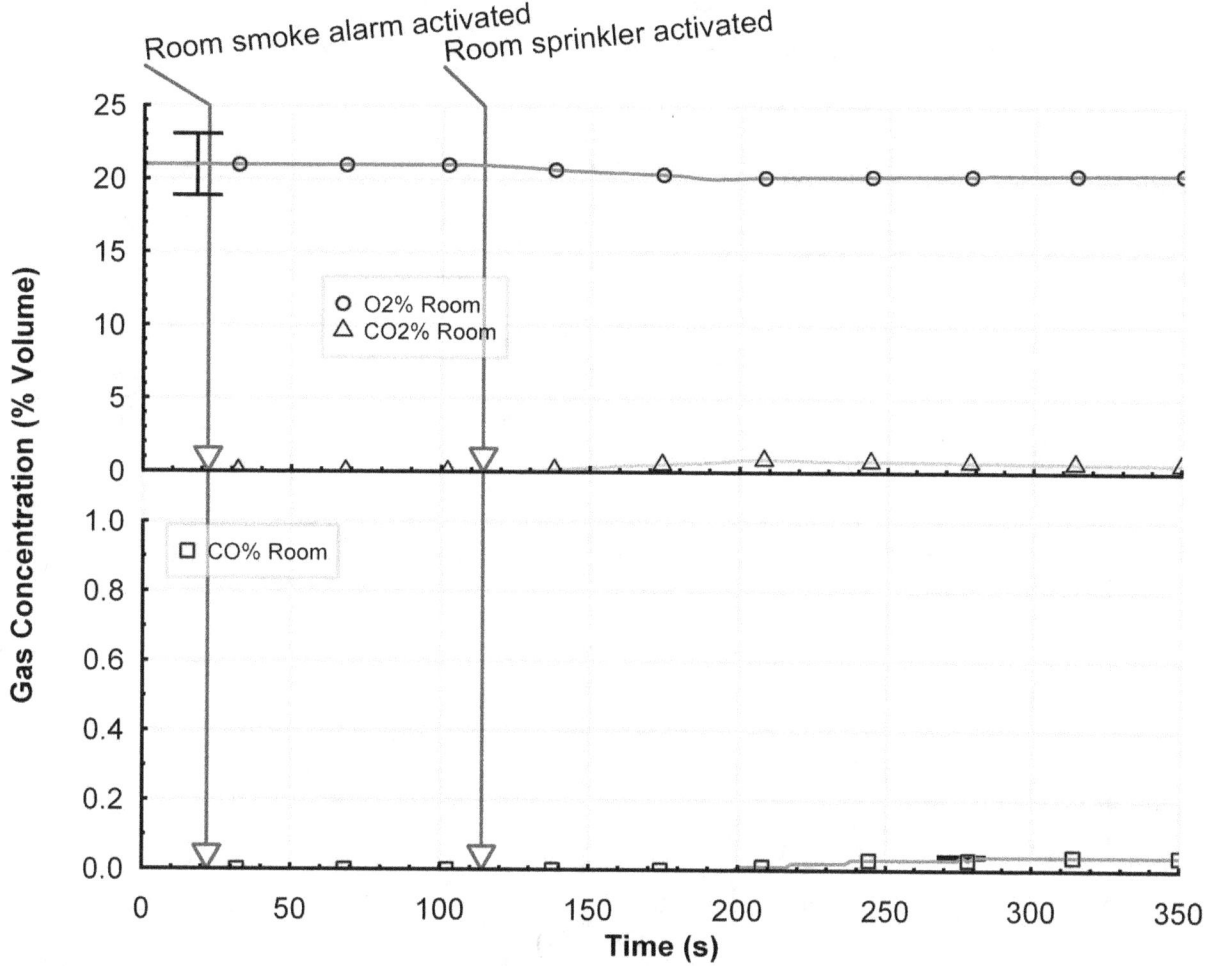

Figure 3.3-6. Relative gas concentrations versus time for the gas sampling at 1.52 m above the floor at location 1 in Experiment 3

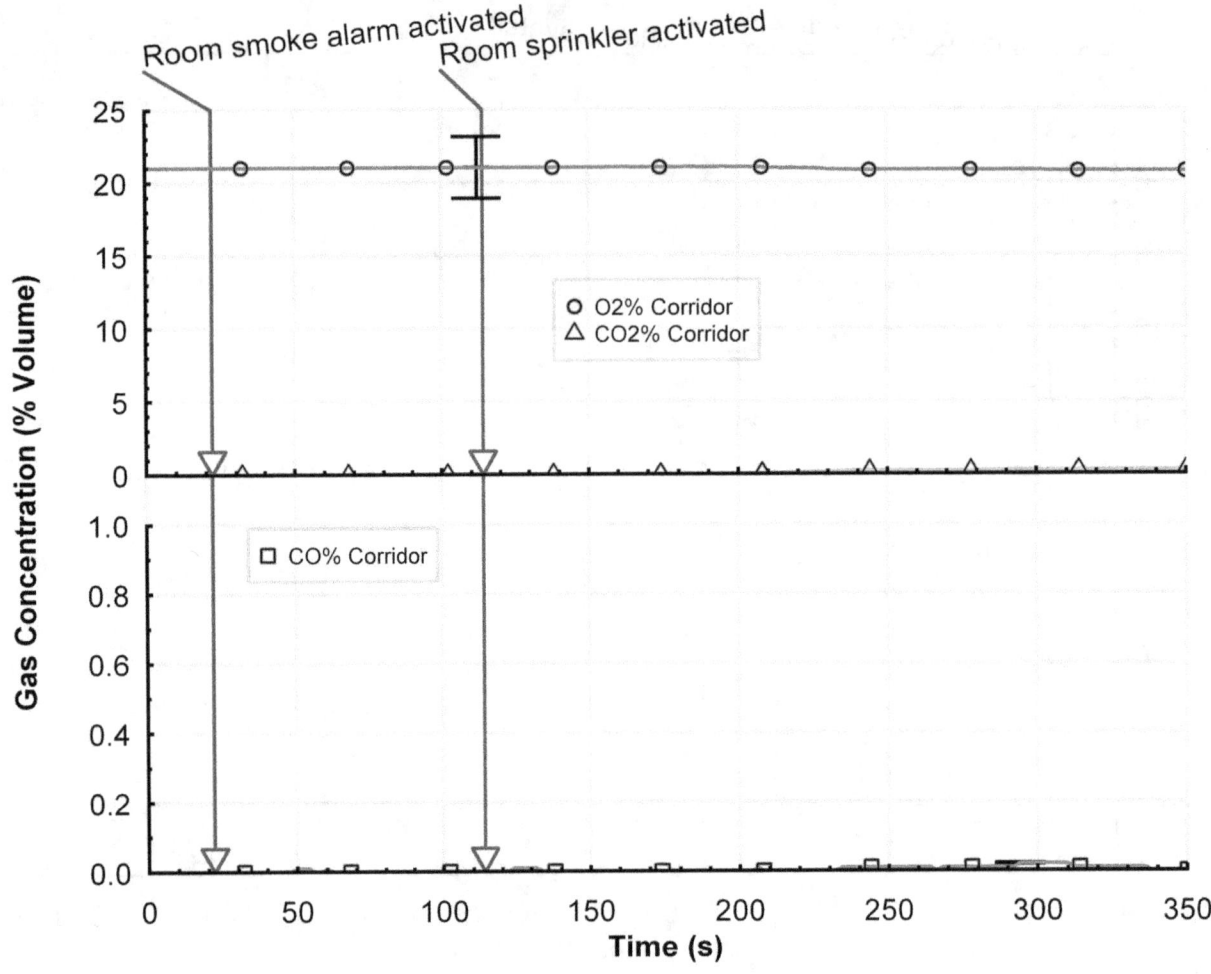

Figure 3.3-7. Relative gas concentrations versus time for the gas sampling at 1.52 m above the floor at location 4 in Experiment 3

3.3.5 Heat Flux Data

The heat flux data from the three sets of gauges positioned in the corridor are shown in Figure 3.3-8. Even with the open door, there was very little increase in heat flux at the gauges locations, as there was only a modest amount of heat discharged into the corridor from the fire room, prior to sprinkler activation. The gauges at location 4, being closest to the open doorway of the fire room, increased on average to approximately 0.25 kW/m^2.

42

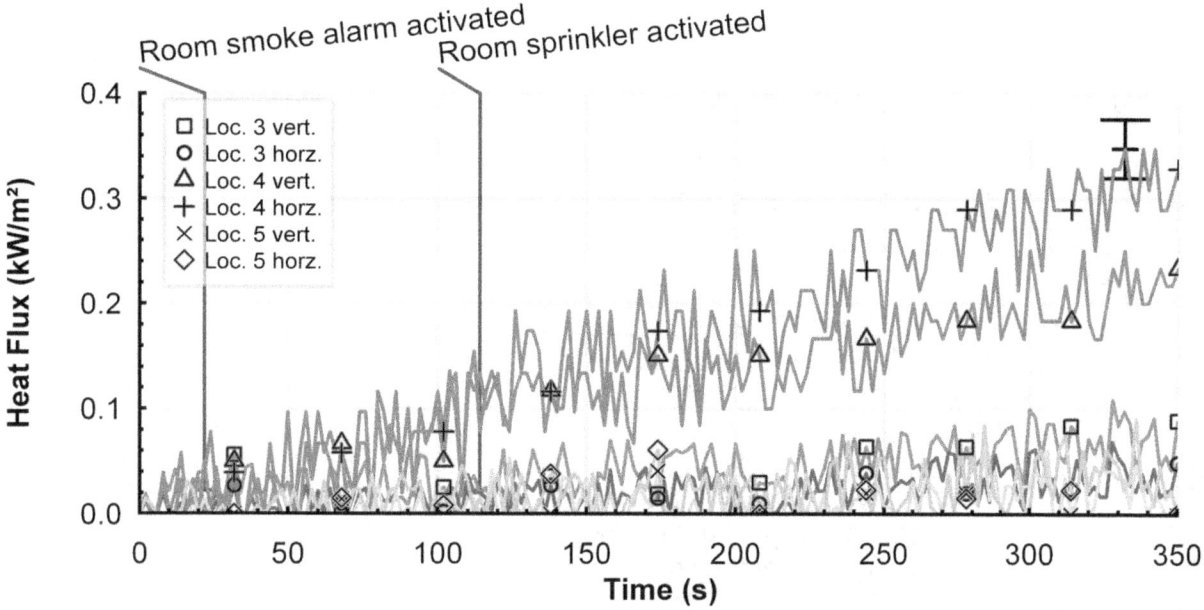

Figure 3.3-8. Heat flux versus time data for the heat flux gauges at locations 3, 4, and 5 in the orientations up (vert.) and sideways (horz.) in Experiment 3

3.4 Experiment 4: Open Door, Non-Sprinklered

The objective of this experiment was examine the fire development and measure the level of hazard produced when there was compromised compartmentation, due to the open door, and no automatic fire suppression system. This experiment also provided a direct comparison with the previous experiment as a means to demonstrate the impact of the automatic sprinkler.

3.4.1 Experiment Timeline

The timeline was developed from observations made during the experiment, review of the video of the experiment, and review of the data. Table 3.4-1 describes the level of fire development in the room. This can be compared with other measurements presented in following sections, such as changes in temperature, gas concentration or fire suppression team response.

Table 3.4-1. Timeline for Experiment Number 4

Time (s)	Observations
0	Ignition
14	Smoke alarm in fire room activated
15	Flames extended to bedding
30	Flames extended to pillow
60	Visible smoke layer started to form
75	Visible smoke started to enter corridor
76	Tell-tale sprinkler in fire room activated
80	Flames extended to desk
90	Flames extended to floor carpeting
100	Flames extended to plastic bookshelf
110	Visible smoke layer started to form in corridor
128	Tell-tale sprinkler in corridor activated
145	Plastic bookshelf falls
202	Right window failure starting
220	Right window failure complete
220	Center window failure starting
225	Fire out of right window
287	Center window failure complete
290	Fire out of center window
319	Left window failure starting
328	Left window failure complete
	Fire out of all three windows
430	Smoke door open, firefighter enter
465	Hose stream suppression
550	Fire extinguished, suppression completed

3.4.2 Smoke Alarm and Sprinkler Activation Times

The smoke alarm activation and sprinkler activation times are given in Table 3.4-2 and Table 3.4-3, respectively. The smoke alarm located in the dorm room activated at 14 s, just as the fire was spreading to the bedding. The temperature at the thermocouple located near the smoke alarm was 45 °C (113 °F) at the time of activation. Due to the open door between the dorm room and the corridor, within 62 s after ignition, the three smoke alarms installed on the ceiling of the corridor activated.

The fire generated enough heat to activate the sprinkler in the dorm room 76 s after ignition. At this point, the temperature adjacent to the sprinkler was approximately 136 °C (277 °F). In this experiment, the sprinkler did not have a water supply, it was installed only to provide an activation time.

Given that there was no suppression from the sprinkler in the dorm room, the fire continued to grow unabated. As a result, the sprinkler located in the corridor was activated at 128 s after ignition, being exposed to a gas temperature of 99 °C (210 °F). These sprinklers had no water supply and were used only to provide an activation time.

Table 3.4-2. Dorm Room Experiment 4, Smoke Alarm Activation Times

Smoke Alarm	Location	Time (s)	Temperature (°C)
1	Dorm Room	14	45
2	West Corridor	60	31
3	Center Corridor	32	29
4	East Corridor	62	32

Table 3.4-3. Dorm Room Experiment 4, Sprinkler Activation Times

Sprinkler Activation	Location	Time (s)	Temperature (°C)
1	Dorm Room	76	136
2	Corridor	128	99

3.4.3 Temperature Data

Temperatures from the 5 thermocouple arrays located in the dorm room (TC arrays 1 and 2) and in the corridor (TC arrays 3 through 5) are presented in Figure 3.4-1 through Figure 3.4-5.

Figure 3.4-1 and Figure 3.4-2 provide the temperature time history for both of the thermocouple arrays located in the dorm room. Both graphs show the steady development of a hot gas layer in the dorm room and rapid progression to a post-flashover condition at approximately 220 s after ignition. At 220 s after ignition, the thermal conditions had transitioned from having a hot layer and a cold layer in the room to having a single well mixed reaction zone with nearly equal temperatures floor to ceiling in excess of 600 °C (1112 °F). This post-flashover burning condition continued until fire fighters began to suppress the fire 465 s after ignition. The firefighters completed their suppression efforts approximately 85 s later, 550 s after ignition.

After the suppression was complete, temperatures in the room remained elevated due to the "stored energy" in concrete ceiling and walls of the room. The temperatures in TC array 1, were all at 150 °C (302 °F) or below given their position near the now open windows. TC array 2 was located approximately 1.1 m (3.5 ft) north of the south wall of the room. After suppression was completed, the temperatures ranged from approximately 350 °C (660 °F) near the ceiling to approximately 150 °C (302 °F) near the floor.

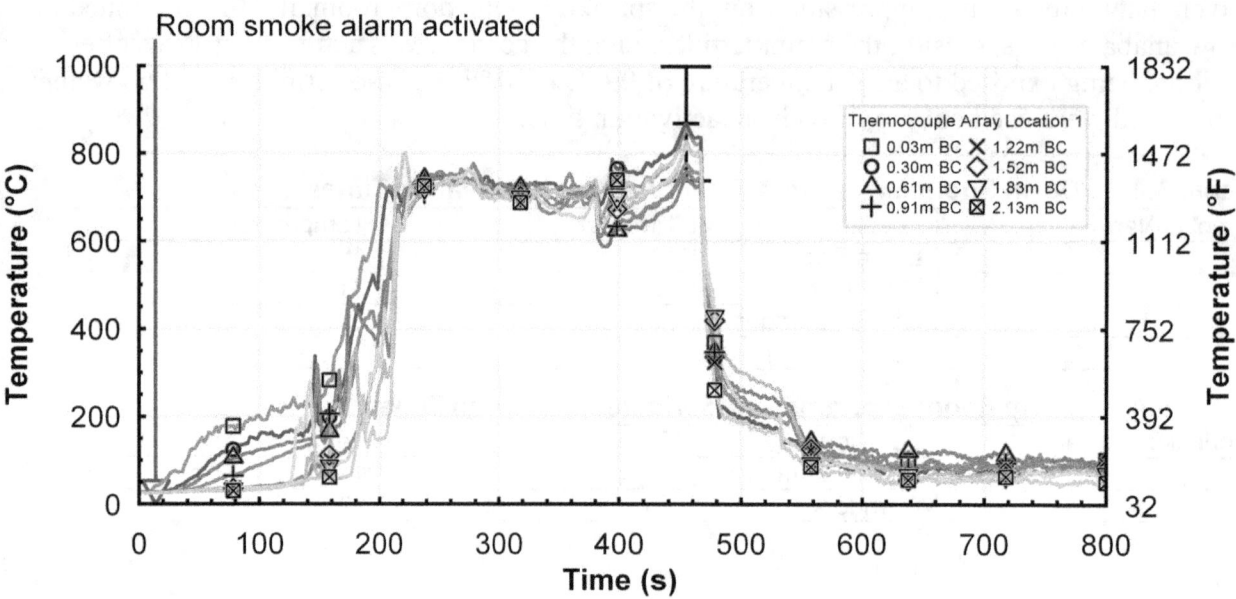

Figure 3.4-1. Temperature versus time data from thermocouple array 1 in Experiment 4, listed by distance below ceiling (BC)

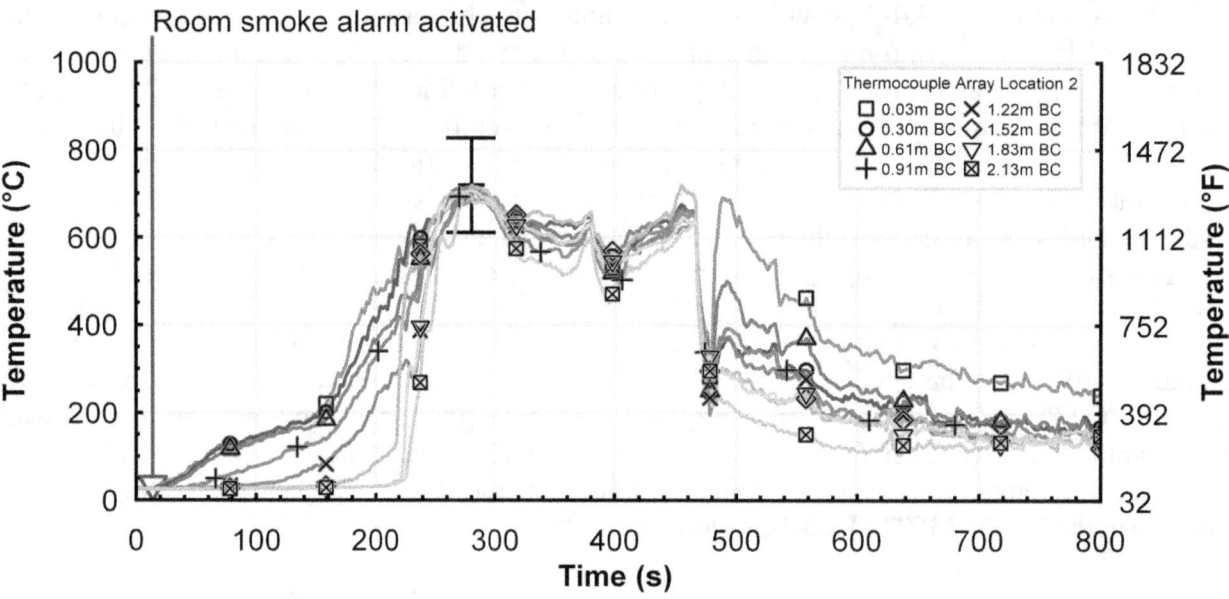

Figure 3.4-2. Temperature versus time data from thermocouple array 2 in Experiment 4, listed by distance below ceiling (BC)

The temperature time histories from the three vertical thermocouple arrays located along the center line of the corridor are given in Figure 3.4-3 through Figure 3.4-5. This experiment had an open door similar to Experiment 3, which allowed hot combustion products to flow into the corridor.

Again TC array 4 was the closest of the three arrays in the corridor to the open doorway of the burning room. As a result, the hot gases reached TC array 4 first and provided the highest temperature increase at that location. The temperature 0.03 m below the ceiling began to increase within 20 s after ignition as shown in Figure 3.4-4. After the dorm room fire transitioned to a post-flashover fire at 220 s after ignition, the peak temperatures near the ceiling in the corridor were on the order of 300 °C (572 °F) until suppression began.

The temperatures from TC array 3 and 5 are shown in Figure 3.4-3 and Figure 3.4-5, respectively. Each of these arrays are positioned 4.6 m (15 ft) away from location 4, although in opposite directions. At both of these positions the temperatures significantly decreased relative to TC array 4. The peak temperatures near the ceiling never exceeded 200 °C (392 °F).

The corridor itself did not transition to flashover, as shown by the thermal gradient with a hot gas layer in the upper portion of the corridor and the near ambient temperatures in the lower portion of the corridor, as shown in Figure 3.4-3 through Figure 3.4-5. This condition began as hot combustion products entered the corridor and continued until suppression. Hence a fire fighter crouching 1.2 m (4.0 ft) or less above the floor as they approached the fire room would be exposed to temperatures of 100 °C (212 °F) or less.

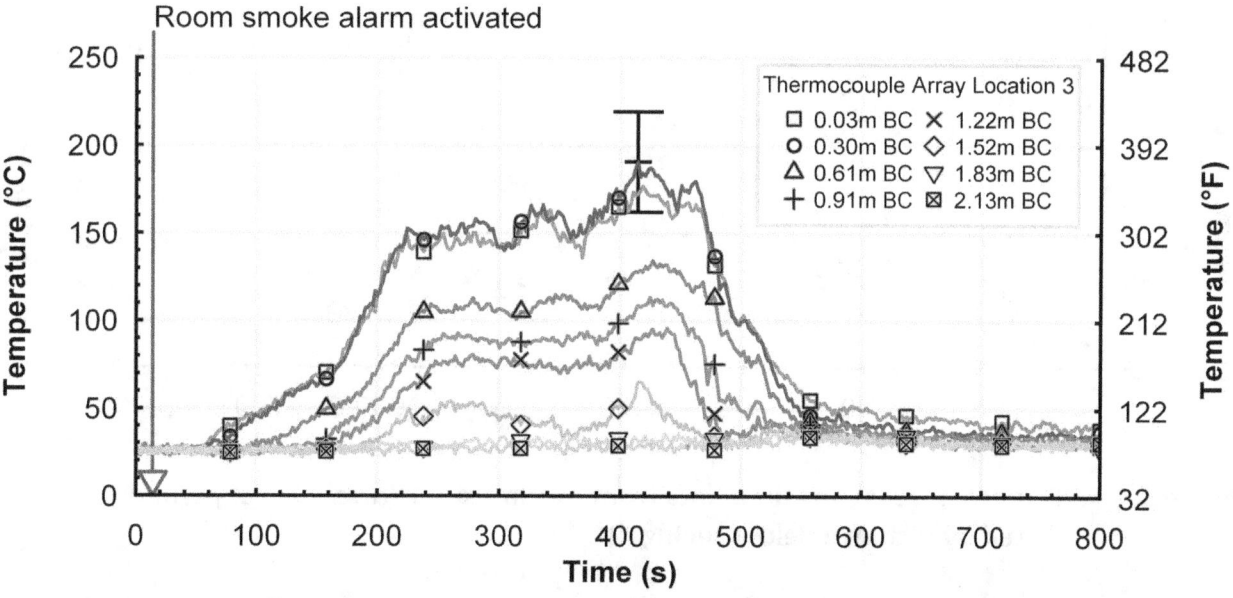

Figure 3.4-3. Temperature versus time data from thermocouple array 3 in Experiment 4, listed by distance below ceiling (BC)

47

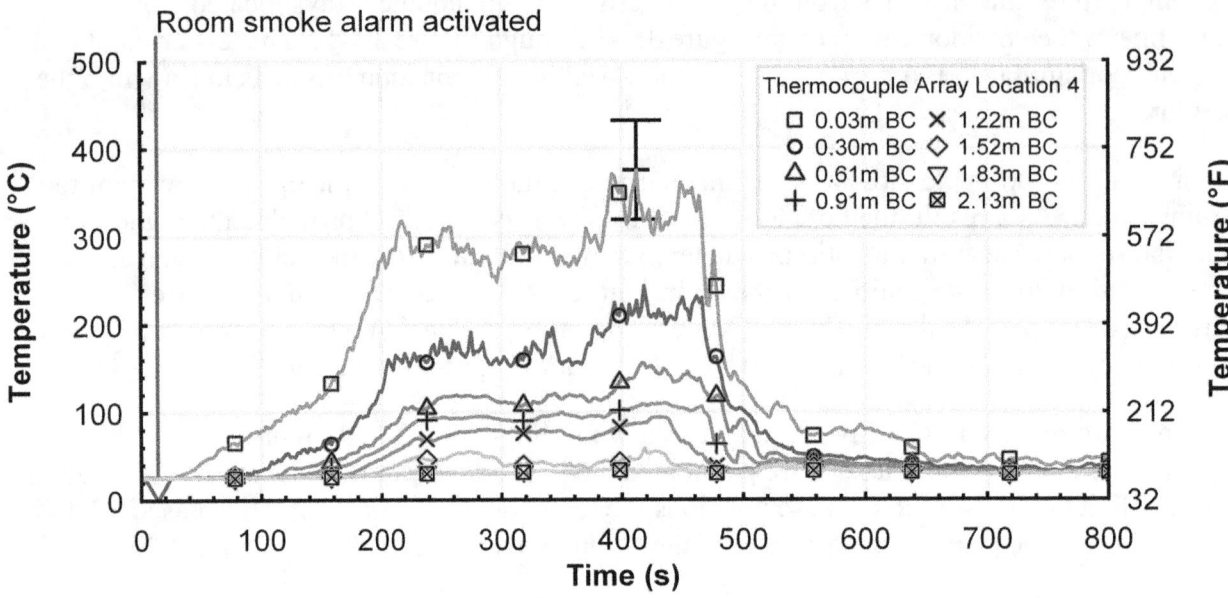

Figure 3.4-4. Temperature versus time data from thermocouple array 4 in Experiment 4, listed by distance below ceiling (BC)

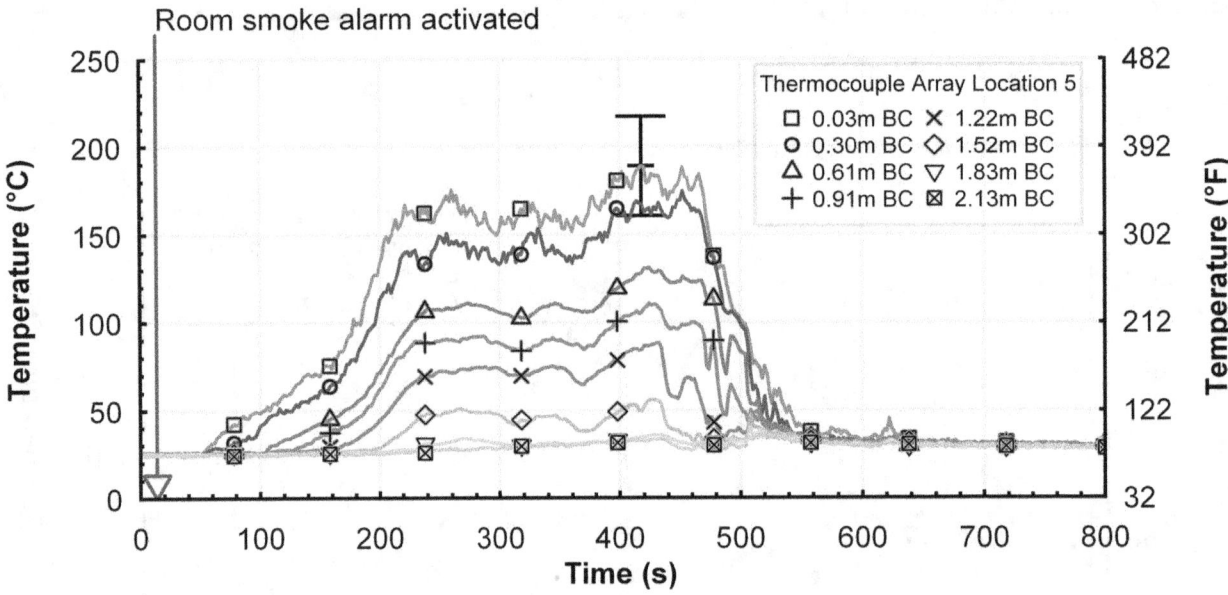

Figure 3.4-5. Temperature versus time data from thermocouple array 5 in Experiment 4, listed by distance below ceiling (BC)

3.4.4 Gas Concentrations

The gas concentrations were sampled in two locations, one in the dorm room adjacent to TC array 1 and one centered in the corridor adjacent to TC array 4; their measured values with respect to time are shown in Figure 3.4-6 and Figure 3.4-7, respectively. Both of the sample locations were positioned 1.52 m (5.0 ft) above the floor. In each figure, the oxygen and carbon dioxide levels are given on the upper graph with a range of 0 % to 25 % by volume and the carbon monoxide measurement is shown on the lower graph which has a range of 0 % to 8 % by volume for the dorm room location and a range of 0 % to 4% by volume for the corridor location.

The oxygen concentration in the dorm room began to decrease at approximately 140 s after ignition, which indicated that the hot gas layer had extended down from the ceiling to the entry level of the sampling tube. The carbon dioxide began to increase noticeably as the oxygen decreased. At approximately 250 s after ignition, the oxygen concentration started to decrease at a faster rate, decreasing from 18 % to approximately 1 % by volume within a 150 s period. During this same period, the carbon dioxide increased to 3 %. At this point in the development of the fire, all of the windows had self-vented and the temperatures in the dorm room had been consistent with post-flashover conditions for the past 180 s.

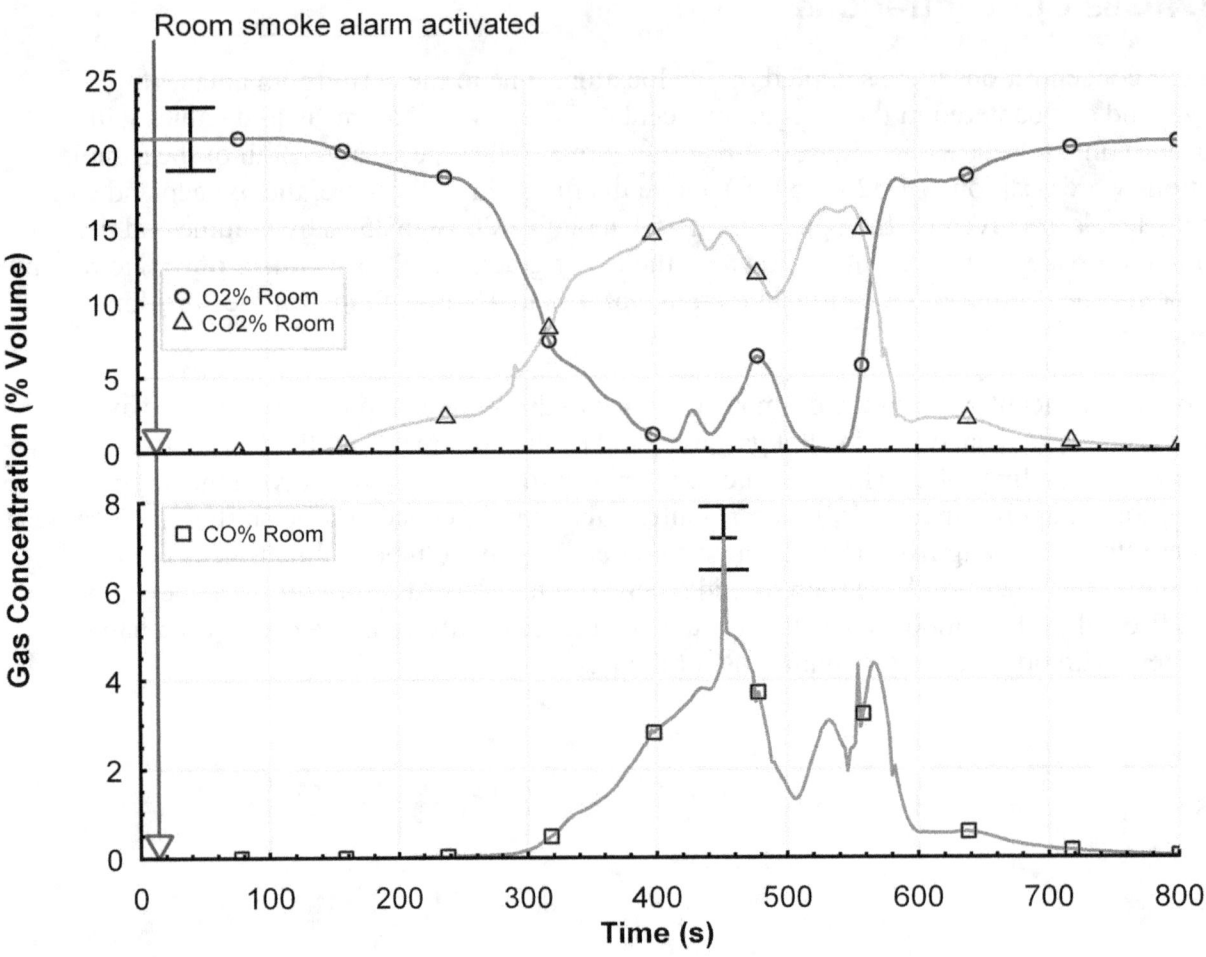

Figure 3.4-6. Relative gas concentrations versus time for the gas sampling at 1.52 m above the floor at location 1 in Experiment 4

Figure 3.4-7 shows the measured gas concentrations from the corridor location. The gas concentrations began to change significantly at approximately 250 s after ignition. The trends of decreasing oxygen, and increasing carbon dioxide and carbon monoxide continued until after fire suppression started. Oxygen concentrations reached a low of approximately 5 %. Carbon dioxide and carbon monoxide reached peak values of 13 % and 3 %, respectively.

The gas concentrations in the dorm room and the corridor returned to near pre-fire by 800 s after ignition.

50

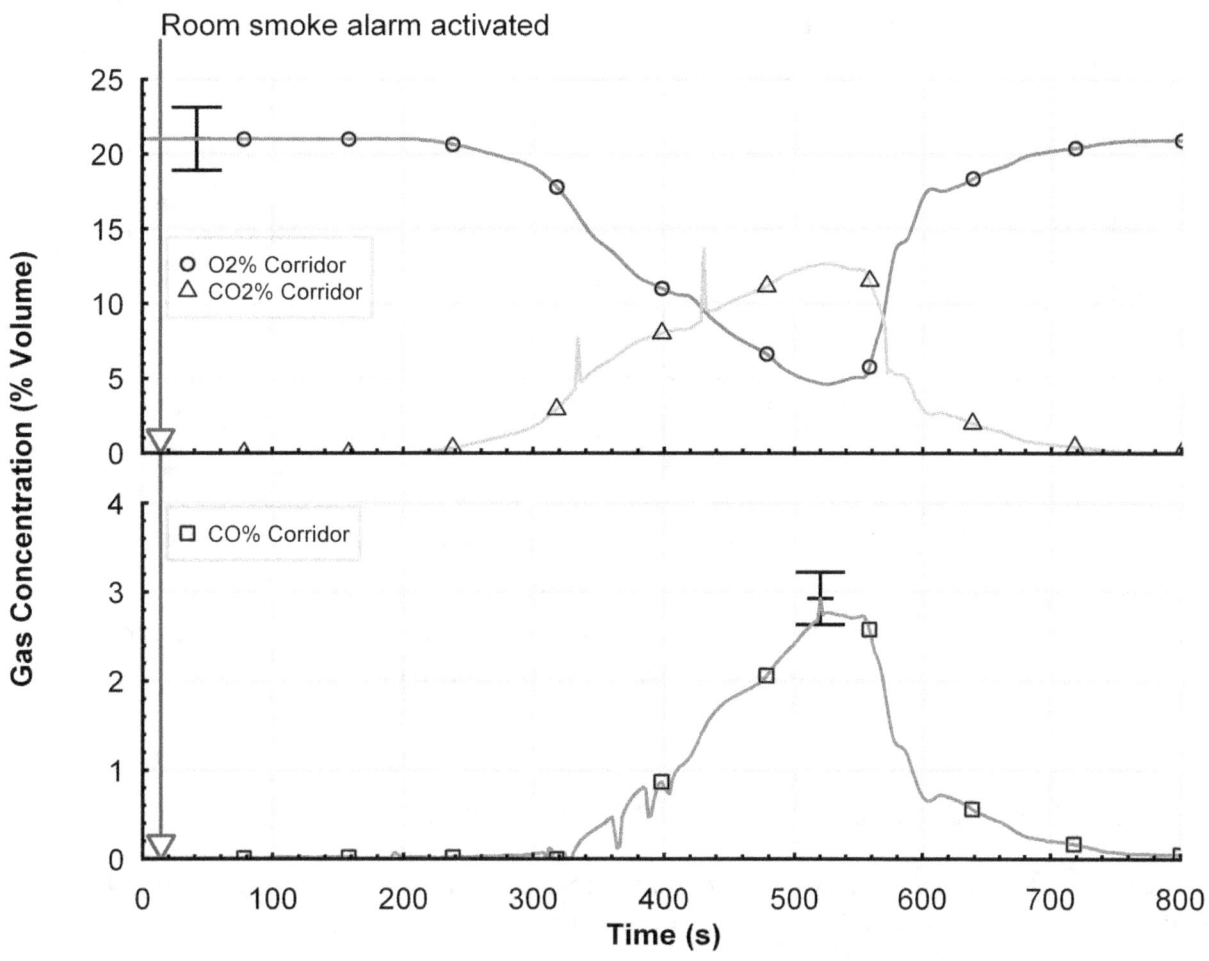

Figure 3.4-7. Relative gas concentrations versus time for the gas sampling at 1.52 m above the floor at location 4 in Experiment 4

3.4.5 Heat Flux Data

The heat flux data from the three sets of gauges positioned in the corridor are displayed in Figure 3.4-8. The gauges at location 4, being closest to the open doorway of the fire room exhibit the largest increase of the three pairs. The peak heat flux was between 2 kW/m^2 and 3 kW/m^2 as measured by the horizontal gauge, pointed toward the open doorway of the fire room. The heat flux values are consistent with the temperature gradient in the corridor as discussed in Section 3.4.3.

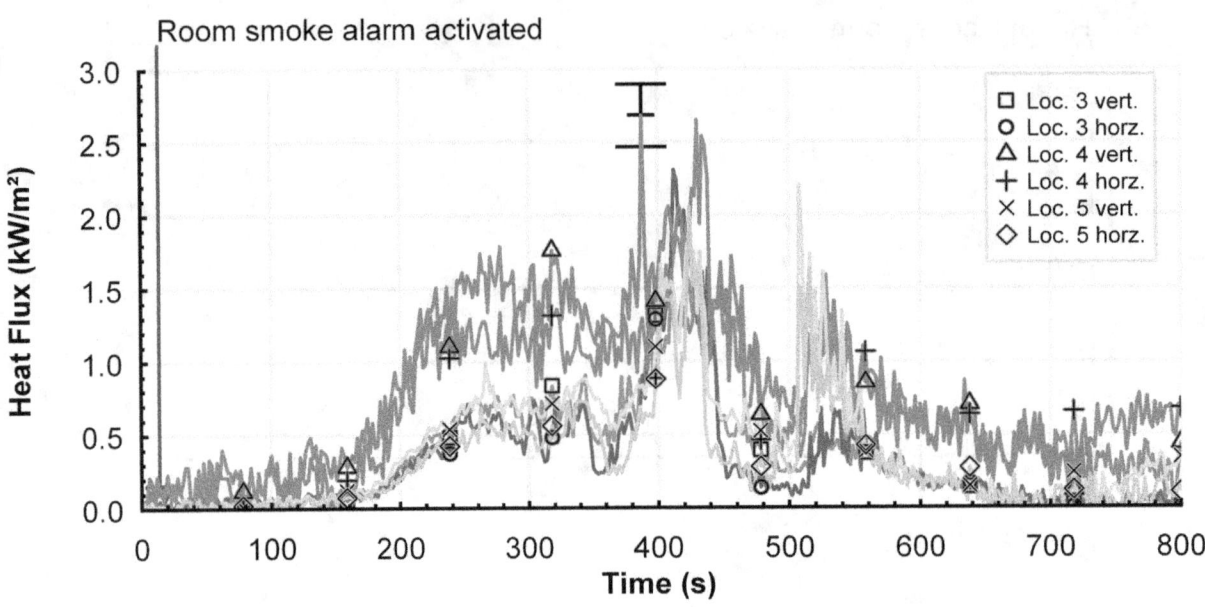

Figure 3.4-8. Heat flux versus time data for the heat flux gauges at locations 3, 4, and 5 in the orientations up (U) and sideways (S) in Experiment 4

3.5 Experiment 5: Open Door, Non-Sprinklered

The objective of this experiment was to replicate Experiment 4 and provide another baseline data set with an open door and no automatic fire suppression.

3.5.1 Experiment Timeline

The timeline was developed from observations made during the experiment, review of the video of the experiment, and review of the data. Table 3.5-1 describes the level of fire development in the room. This can be compared with other measurements presented in following sections, such as changes in temperature, gas concentration or fire suppression team response.

Table 3.5-1. Timeline for Experiment Number 5

Time (s)	Observations
0	Ignition
26	Smoke alarm in fire room activated
25	Visible smoke layer started to form
35	Flames extended to bedding
40	Smoke started to enter the corridor
45	Flames extended to pillow
70	Smoke layer started to form in corridor
110	Tell-tale sprinkler in fire room activated
165	Flames extended to desk
180	Plastic bookshelf falls
191	Right window failure starting (glass cracking)
224	Tell-tale sprinkler in corridor activated
246	Smoke to the floor
247	Center window failure starting
290	Flames out of small openings in right window glass
305	Left window failure starting
331	Right window failure complete
405	Flames out of openings near top of left window glass
417	Center window failure complete
426	Left window failure complete
501	Door open
570	Start suppression

3.5.2 Smoke Alarm and Sprinkler Activation Times

The smoke alarm activation and sprinkler activation times are given in Table 3.5-2 and Table 3.5-3 respectively. The smoke alarm located in the dorm room activated at 26 s. The temperature at the thermocouple located near the smoke alarm was 31 °C (88 °F) at the time of activation. Due to the open door between the dorm room and the corridor, within 98 s after ignition all three smoke alarms installed on the ceiling of the corridor had activated. The time until activation for the most remote smoke alarm was 36 s longer than in Experiment 4. This was due in part to the burn room being at the west end of the corridor instead of near the middle as in the previous experiment.

The fire generated enough heat to activate the sprinkler in the dorm room at 110 s after ignition. At this point the temperature adjacent to the sprinkler was approximately 82 °C (180 °F). In this experiment, the sprinkler did not have a water supply, it was installed only to provide an activation time.

Given that there was no suppression from the sprinkler in the dorm room, the fire continued to grow. As a result, the sprinkler located in the corridor at location 4 was activated at 224 s after ignition. The gas temperature near the sprinkler at the time of activation was 125 °C (257 °F). This sprinkler was also intended to only provide an activation time.

Table 3.5-2. Dorm Room Experiment 5, Smoke Alarm Activation Times

Smoke Alarm	Location	Time (s)	Temperature (°C)
1	Dorm Room	26	31
2	West Corridor	62	30
3	Center Corridor	80	31
4	East Corridor	98	31

Table 3.5-3. Dorm Room Experiment 5, Sprinkler Activation Times

Sprinkler Activation	Location	Time (s)	Temperature (°C)
1	Dorm Room	110	82
2	Corridor	224	125

3.5.3 Temperature Data

Temperatures from the 5 thermocouple arrays located in the dorm room (TC arrays 1 and 2) and in the corridor (TC arrays 3 through 5) are presented in Figure 3.5-1 through Figure 3.5-5.

Figure 3.5-1 and Figure 3.5-2 provide the temperature time history for both of the thermocouple arrays located in the dorm room. Both graphs show the steady development of a hot gas layer in the dorm room reaching temperatures of approximately 200 °C (392 °F) by 180 s after ignition.

After 180 s, the temperature gradients in the two different areas of the dorm room diverge. The temperatures near the ceiling at TC array 1, near the area of origin, increase to approximately 800 °C (1470 °F) at 240 s after ignition. At the same time the temperature 1.52 m (5 ft) below the ceiling was 100 °C (212 °F). There was also a thermocouple in the array 1.83 m (6 ft) below the ceiling, whose temperature did not follow the trends of the thermocouples either above or below it. A plastic crate fell and pushed TC array 1 toward the burning bed, and the TC at the 1.83 m position was pushed into the burning bed at that point. This was the only TC affected.

During this same time, 240 s after ignition, the temperatures at TC array 2 reached a plateau of approximately 580 °C (1076 °F) near the ceiling. At the same time, the temperatures near the floor of this position were still less than 100 °C (212 °F).

At 330 s after ignition, temperatures in the hot layer began to decrease at both TC array positions in the dorm room. The right window pane had completely vented open by this time and was allowing heat to leave the room. About 40 s, later the temperatures began to increase again. The temperatures in the dorm room increased, transitioning through flashover at approximately 440 s, based on all of temperature measurements from TC array 1 exceeding 600 °C (1112 °F) at that time. The temperatures at TC array 2 showed a similar trend, although it occurred about 20 s later. These temperatures continued until the fire fighters began suppression at 570 s after ignition. Hose stream application resulted in a significant and rapid decrease in temperature throughout the dorm room.

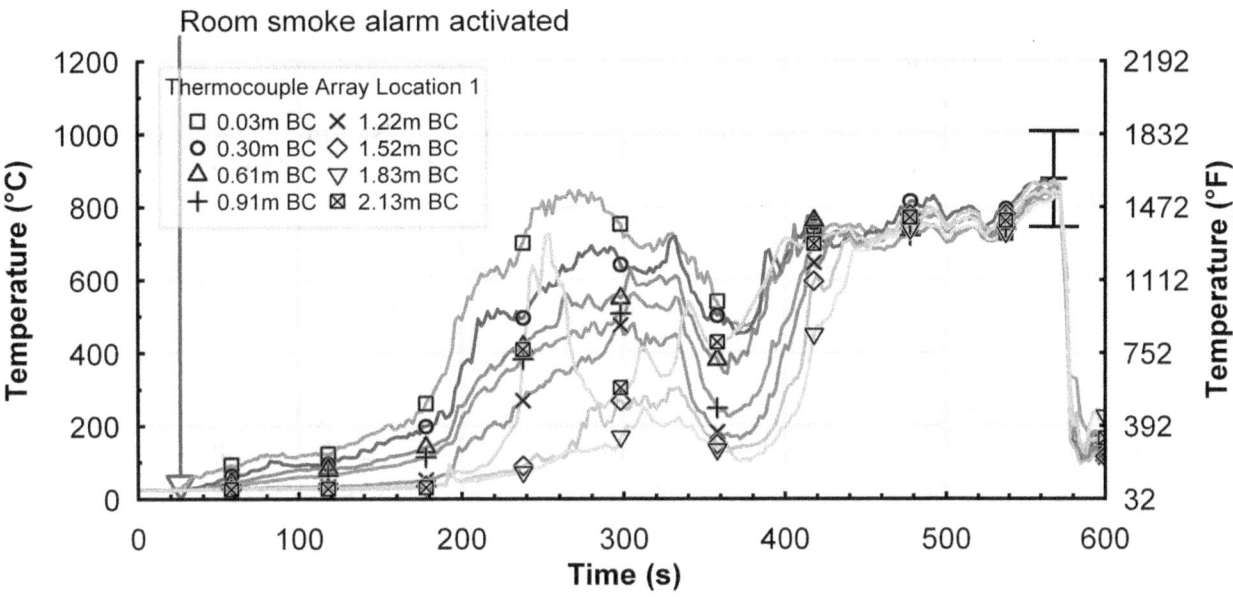

Figure 3.5-1. Temperature versus time data from thermocouple array 1 in Experiment 5, listed by distance below ceiling (BC)

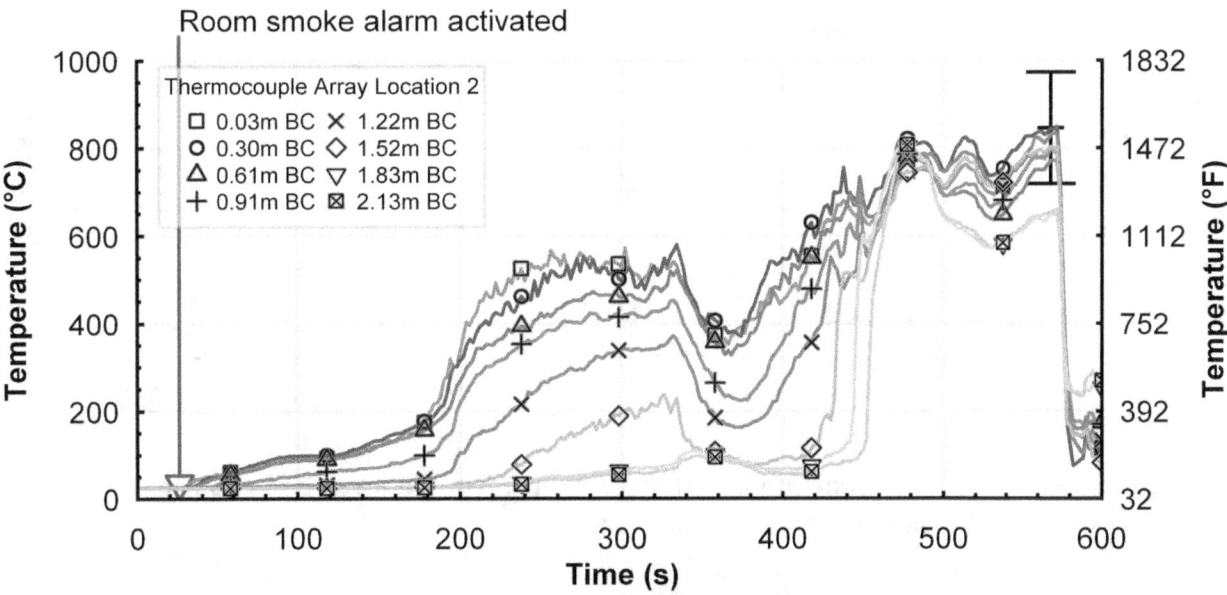

Figure 3.5-2. Temperature versus time data from thermocouple array 2 in Experiment 5, listed by distance below ceiling (BC)

The temperature time histories from the three vertical thermocouple arrays located along the center line of the corridor are given in Figure 3.5-3 through Figure 3.5-5. This experiment had an open door similar to Experiment 3 and 4, which allowed hot combustion products to flow into the corridor. However in this experiment, TC array 3 was closest to the burn room, while TC array 5 was the most remote.

The temperature trends in the corridor follow those from the dorm room in that the temperatures increased, decreased, and then increased again until fire fighting activities commenced with the opening of the smoke door leading into the corridor at 500 s.

The peak temperatures in the corridor were located 0.03m (1 in) below the ceiling for each of the TC arrays in the corridor. The peak temperatures were approximately 230 °C (446 °F), 170 °C (338 °F), and 135 °C (275 °F) at locations 3, 4, and 5 respectively.

As in Experiment 4, the corridor maintained a two layer thermal environment even though the dorm room flashed over. Figure 3.5-3 through Figure 3.5-5 show that throughout the experiment, temperatures 1.22 m (4 ft) below the ceiling or lower never exceeded 100 °C (212 °F).

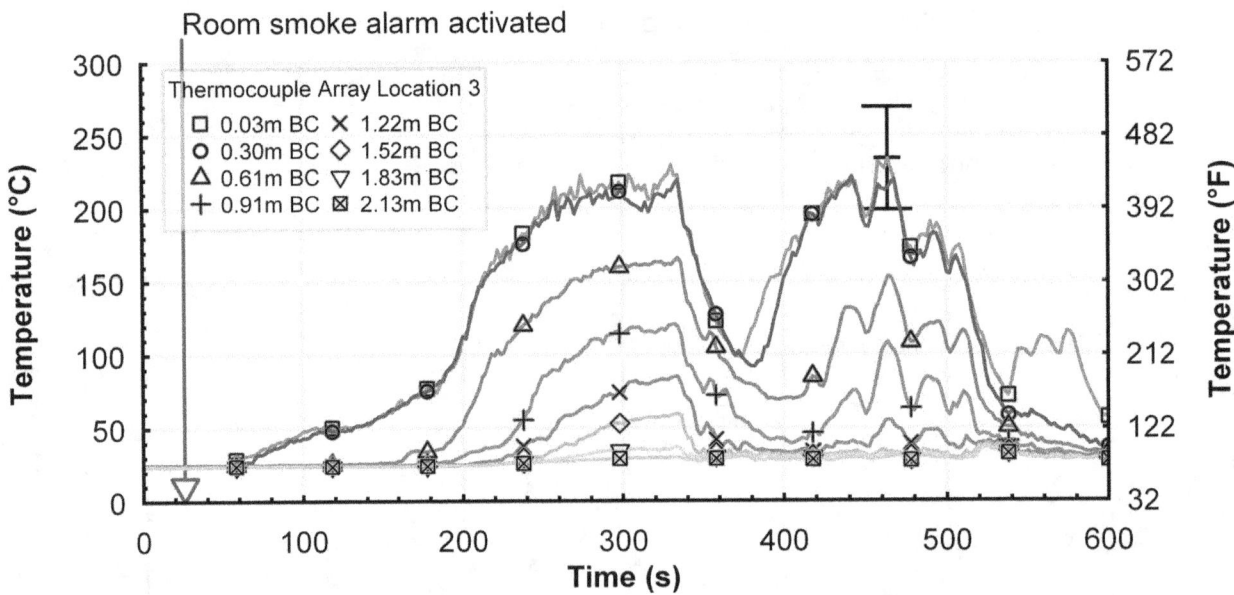

Figure 3.5-3. Temperature versus time data from thermocouple array 3 in Experiment 5, listed by distance below ceiling (BC)

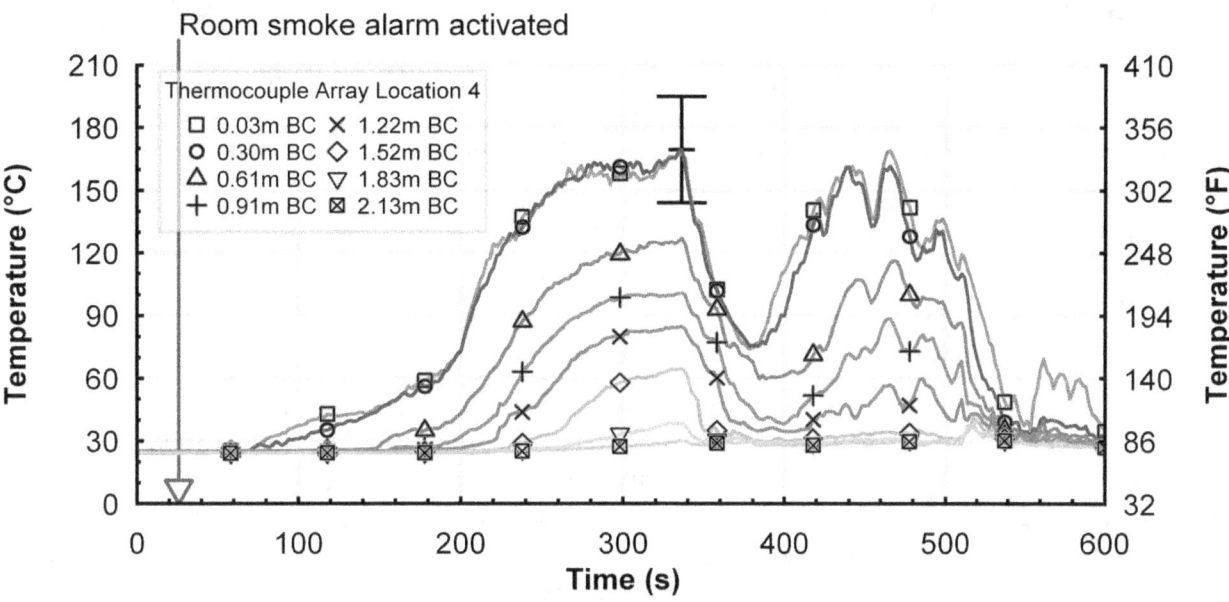

Figure 3.5-4. Temperature versus time data from thermocouple array 4 in Experiment 5, listed by distance below ceiling (BC)

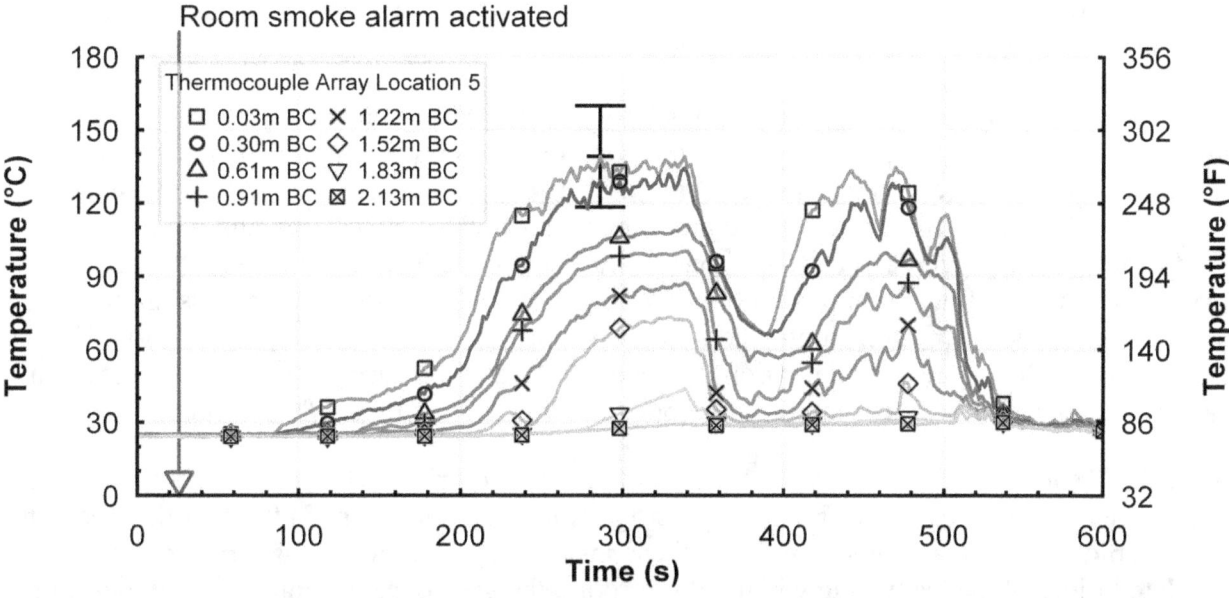

Figure 3.5-5. Temperature versus time data from thermocouple array 5 in Experiment 5, listed by distance below ceiling (BC)

3.5.4 Gas Concentrations

The gas concentrations were sampled in two locations, one in the dorm room adjacent to TC array 1 and one centered in the corridor adjacent to TC array 4; their measured values with respect to time are shown Figure 3.5-6 and Figure 3.5-7 respectively. Both of the sample locations were positioned 1.52 m (5.0 ft) above the floor. In each figure, the oxygen and carbon dioxide levels are given on the upper graph with a range of 0 % to 25 % by volume and the carbon monoxide measurement is shown on the lower graph which has a range of 0 % to 8 % by volume for the dorm room location and a range of 0 % to 2 % by volume for the corridor location.

In the dorm room, the oxygen concentration began to decrease at approximately 120 s after ignition; this occurred in conjunction with the increase in carbon dioxide and carbon monoxide. The rates of change for the gas concentrations are nearly linear until 280 s after ignition, when the rate of change increases. For the interval between 120 s and 280 s, the oxygen concentration decreased approximately 3 % and carbon dioxide increased by approximately 3 %. In absolute terms, the rate of change for both species during this 160 s period was approximately 1.1 % per min. During the next minute, from 280 s to 340 s, the oxygen concentration decreased by 10 %, the carbon dioxide increased by 7 % and the carbon monoxide had increased by 1 %.

The oxygen concentration leveled out at about 6 % between 380 s and 410 s after ignition. This was the period when the temperatures in the dorm room are increasing again, after the opening of one of the three window sections. During this same period, the carbon dioxide increased to a steady level of 11 % by volume and the carbon monoxide reading increased to more than 2 %.

As the temperature in the dorm room continued to increase during the period from 410 s after ignition to flashover at 440 s, the oxygen and carbon dioxide concentrations incurred two rapid transitions with the largest change being an increase of 6 % for the oxygen and about a 4 % decrease for the carbon monoxide. Once all of the windows had self-vented and the fire was in post-flashover, the oxygen and carbon dioxide continued to oscillate for the next 80 s. For the oxygen, the range of fluctuation was between 5 % and 10 %. Just prior to suppression, the temperatures in the dorm room continued to increase, peaking in excess of 820 °C (1500 °F). At that same time, the oxygen concentration dropped to 0 % and the carbon dioxide concentration peaked at 18 % by volume and the carbon monoxide increased to over 6 % by volume.

Figure 3.5-7 shows the gas concentrations measured in the corridor. The gas concentrations did not significantly change for the first 300 s after ignition, then conditions in the corridor began to exhibit rapid change in a manner similar to those inside the dorm room. Just prior to the complete failure of the last window in the dorm room, the gas concentrations reached their pre-suppression peak values. At 420 s after ignition, the oxygen dipped just below 10 %, the carbon dioxide exceeded 8 %, and the carbon monoxide reached 1 %.

After the dorm room windows had completely vented open, the flow path for the fire consisted of drawing fresh air from the building and exhausting the majority of the hot gases out of the window openings. This flow had the impact of increased oxygen concentration and decreased concentrations of carbon dioxide and carbon monoxide. This trend continued until the fire fighters entered the building and suppression activities began.

Although not shown in the figures, the gas concentrations in the dorm room and in the corridor had returned to near pre-fire conditions by 800 s after ignition.

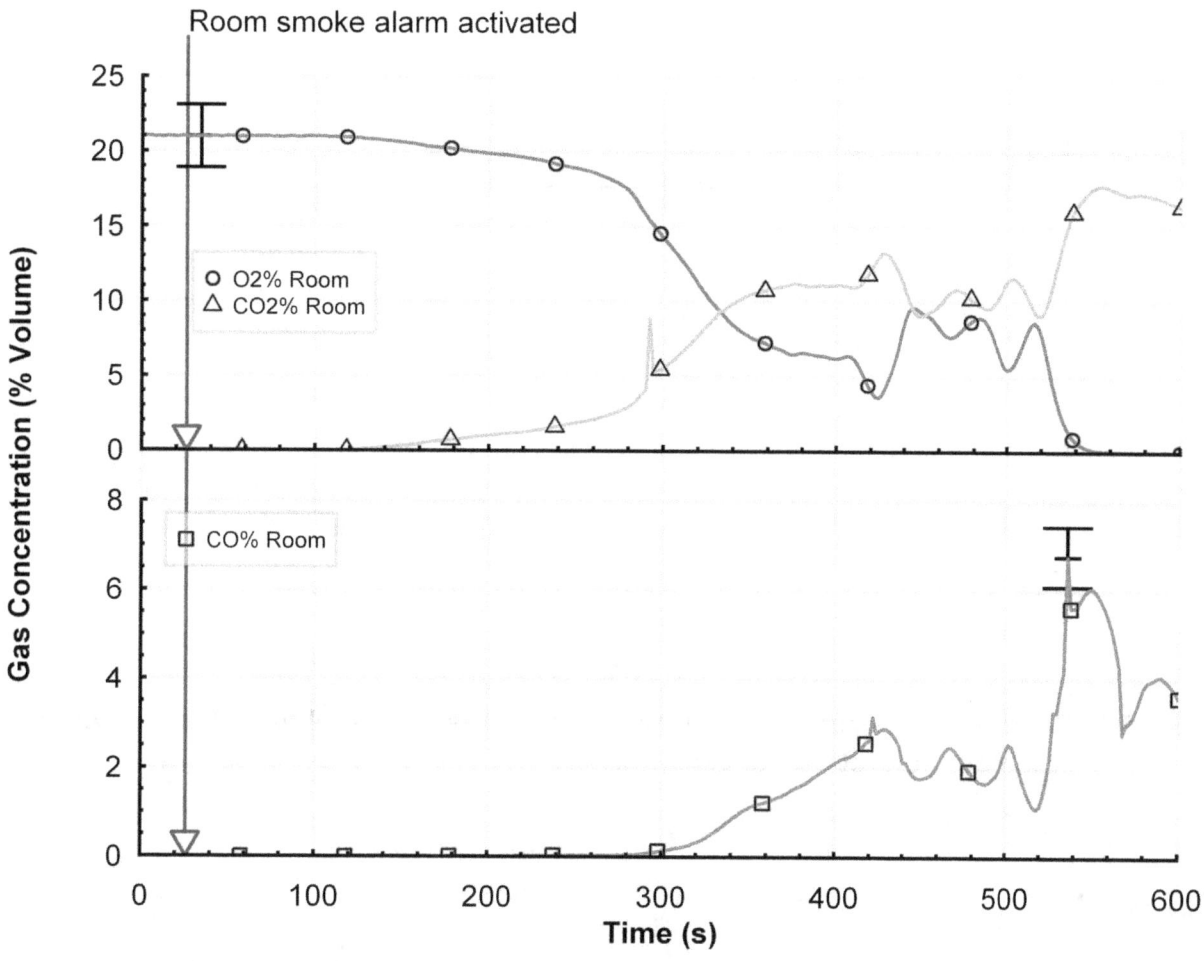

Figure 3.5-6. Relative gas concentrations versus time for the gas sampling at 1.52 m above the floor at location 1 in Experiment 5

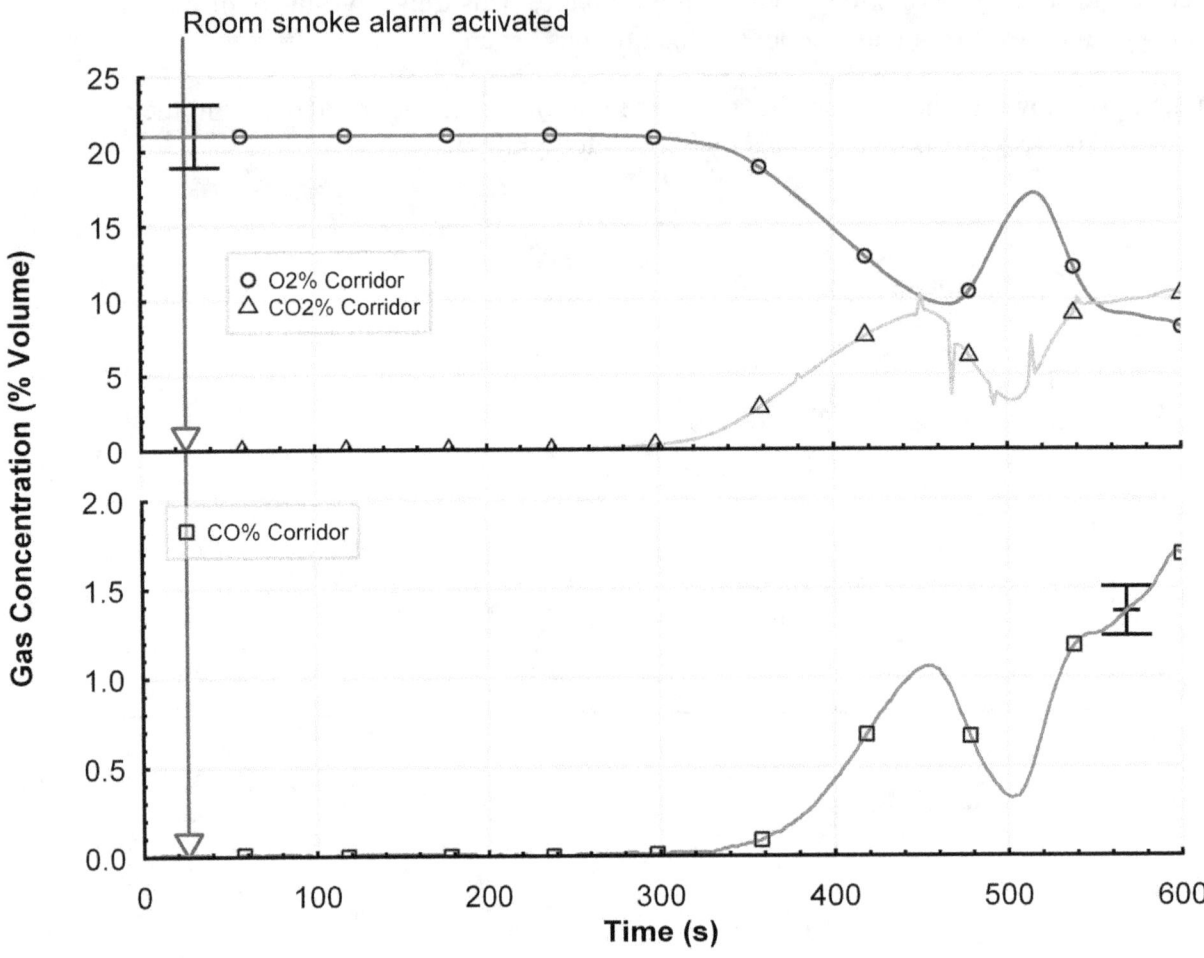

Figure 3.5-7. Relative gas concentrations versus time for the gas sampling at 1.52 m above the floor at location 4 in Experiment 5

3.5.5 Heat Flux Data

The heat flux data from the three sets of gauges positioned in the corridor are displayed in Figure 3.5-8. The gauges at location 3, being closest to the open doorway of the fire room exhibited the largest increase of the three pairs. The peak heat flux was between 2 kW/m^2 and 3 kW/m^2 as measured by the horizontal gauge, pointed toward the open doorway of the fire room. This was consistent with the heat flux results from Experiment 4.

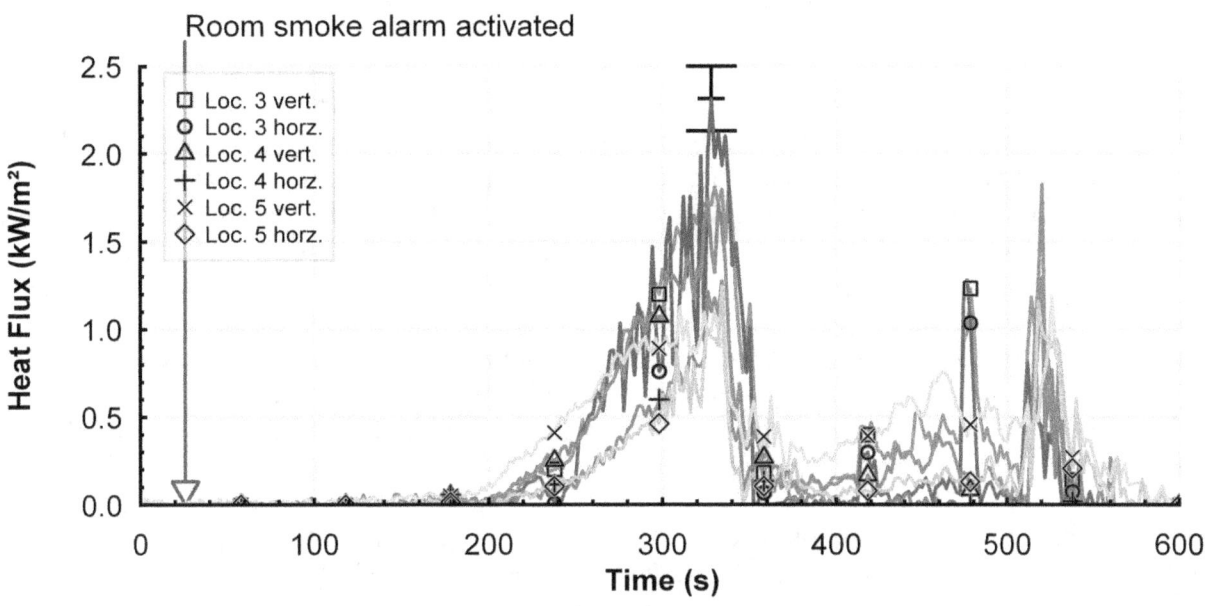

Figure 3.5-8. Heat flux versus time data for the heat flux gauges at locations 3, 4, and 5 in the orientations up (vert.) and sideways (horz.) in Experiment 5

61

4 Discussion

In order to examine the potential impact of the effectiveness of compartmentation or an automatic fire sprinkler system, the discussion will begin with benchmark values for thermal injury and incapacitation due to inhalation of combustion products. The life safety hazards generated by a fire include: temperature, toxic gases, and loss of visibility. In these experiments, only quantitative measures of heat and a few toxic gases were made and some qualitative measures of visibility were made with video cameras.

Heat can be transferred by conduction, convection and radiation. Burn injuries caused by the combustion products (smoke) can be caused by convection and/or radiant heat transfer. Frequently the hazard related to heat transfer from the environment to the body is simplified as an exposure temperature for a prescribed duration.

As presented in the SFPE Handbook of Fire Protection Engineering, estimated limits for tenability due to convected heat suggest a thermal tolerance of 120 °C (248 °F). Above this limit, the onset of pain is rapid and burns can develop within a few minutes or less. The estimated tenability limit due to heat flux is 2.5 kW/m^2. At this level, the time to burn unprotected skin is 20 s or less [13].

These limits are not absolute limits since clothing, humidity, skin composition etc, can mitigate or exacerbate the impact of the thermal energy for a given heat level and exposure time. These values are used as bench marks for the discussion presented here. It is also important to note that as the fire grew, the temperatures in some areas of the dorm room and corridor increased rapidly and quickly exceeded the benchmark tenability thresholds, making concerns over uncertainty in the tenability limits a minor point.

Based on individual exposures, incapacitation can occur due to low oxygen (10 % to 13 %) or high level of carbon dioxide (7 % to 8 %) or carbon monoxide (0.6 % to 0.8 %) after approximately 5 minutes of exposure [13]. Lower concentrations of oxygen and higher concentrations of carbon dioxide and carbon monoxide, or the synergistic effect of all three combustion products would result in incapacitation in a shorter time. For purposes of this analysis, the following values will be considered the limits of untenablity: oxygen < 10 %, carbon dioxide > 8 %, and carbon monoxide > 1 %.

In the sections that follow, comparisons between the experiments are made by examining the temperatures and gas concentrations at approximately 1.5 m (5 ft) above the floor. This elevation can be considered the face height of a typical, standing person.

4.1 Closed Door Experiments

In these experiments, the door between the dorm room and the corridor remained closed throughout the experiments. This had the impact of preventing significant amounts of heat or combustion products from entering the corridor. The conditions in the corridor in both experiments remained tenable.

The post fire damage in both dorm rooms was similar, as shown in Figure 4.1-1. In the non-sprinklered experiment, the fire self-extinguished due to low oxygen concentration, while in the other dorm room with a closed door, the water from the sprinkler extinguished the fire.

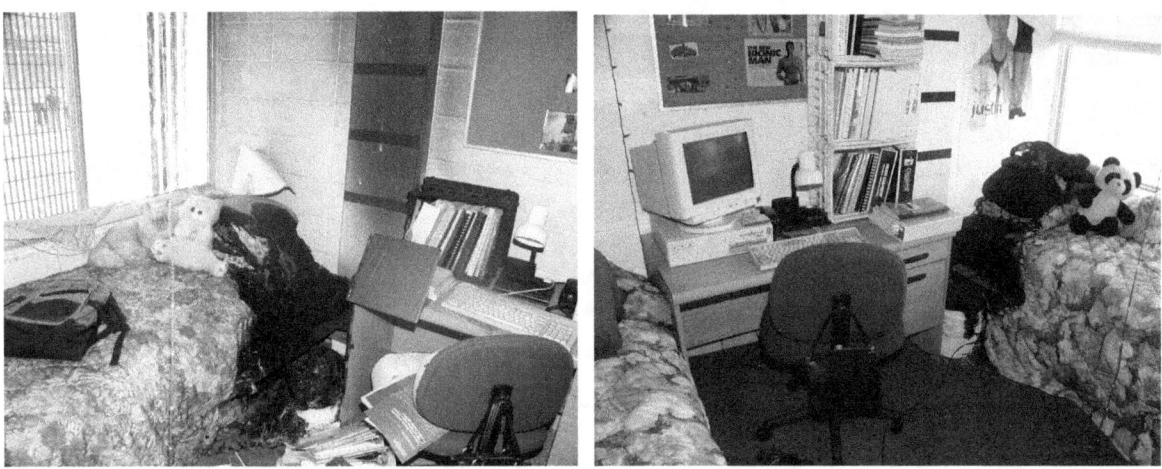

Figure 4.1-1. Post fire photographs of the dorm room for Experiment 1 (non-sprinklered) on the left and Experiment 2 (sprinklered) on the right.

Even though the fire damage was similar between Experiment 1 and Experiment 2, the tenability conditions in the dorm rooms were quite different. Figure 4.1-2 shows the measured temperatures from the TC array 1 location in Experiment 1 and Experiment 2. The temperatures in both experiments increased after ignition and followed similar trends until the sprinkler in Experiment 2 activated at 100 s. The temperatures significantly diverged at this point, with the temperatures in the non-sprinklered experiment surpassing untenable levels at approximately 150 s after ignition. The fire in Experiment 2 was automatically extinguished 125 s after ignition.

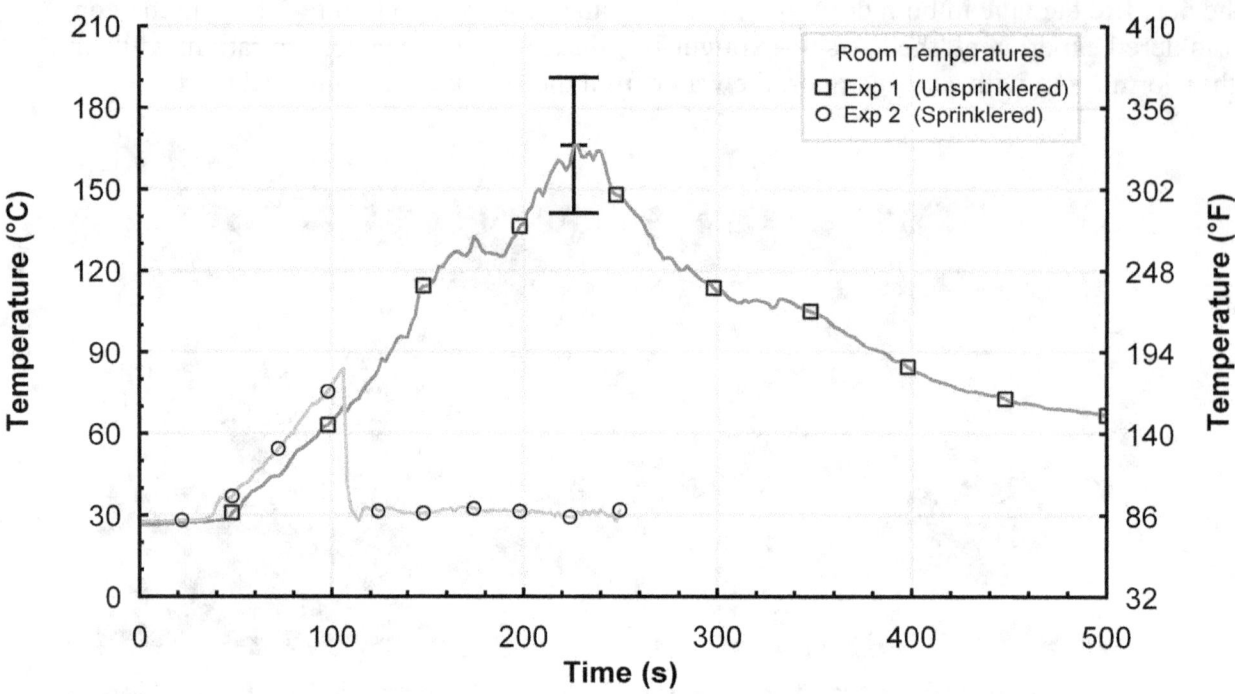

Figure 4.1-2. Temperature versus time data from thermocouple array 1 from Experiment 1 and 2 at approximately 1.5 m (5 ft) above the floor.

Figure 4.1-3 through Figure 4.1-5 show the comparison of oxygen, carbon dioxide and carbon monoxide for the two experiments. In both experiments, the measurements showed decreased oxygen levels and increased amounts of carbon dioxide and carbon monoxide. In the sprinklered experiment, this occurred after sprinkler activation due to the cooling and mixing of the combustion products in the hot gas layer. The gas concentration levels in the sprinklered experiment as the fire was being suppressed were approximately five times less than the peak values in the non-sprinklered experiment and remained within tenable limits throughout the experiment.

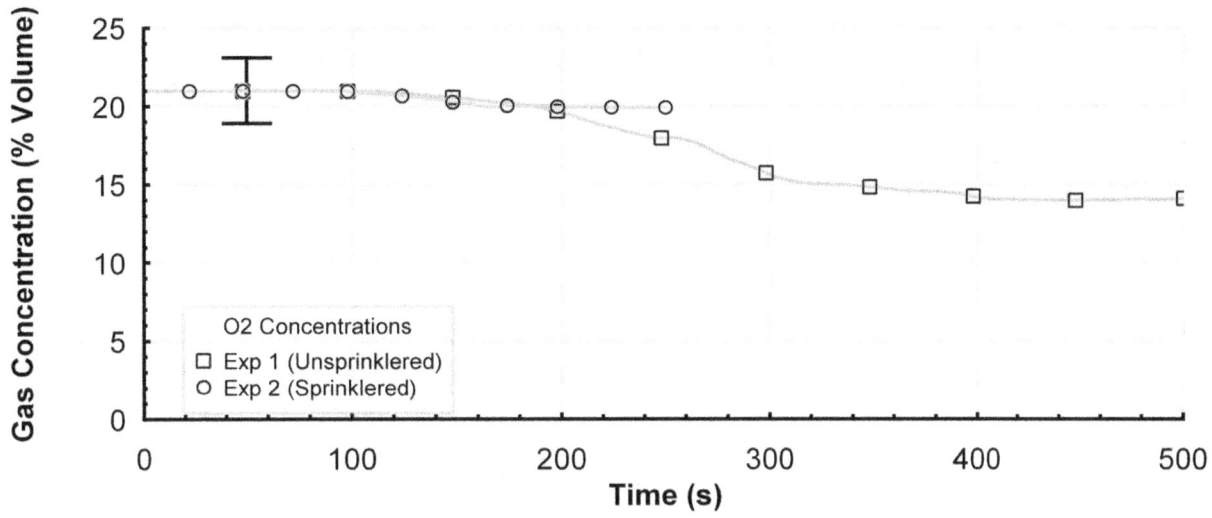

Figure 4.1-3. Oxygen concentrations versus time in the dorm room from Experiment 1 and 2 at approximately 1.5 m (5 ft) above the floor.

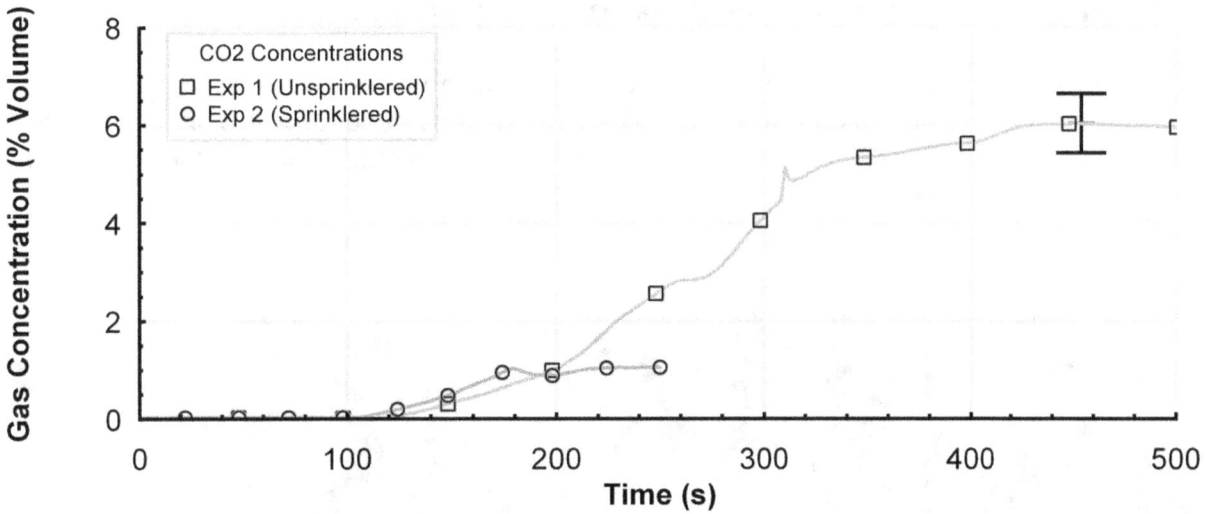

Figure 4.1-4. Carbon dioxide concentrations versus time in the dorm room from Experiment 1 and 2 at approximately 1.5 m (5 ft) above the floor.

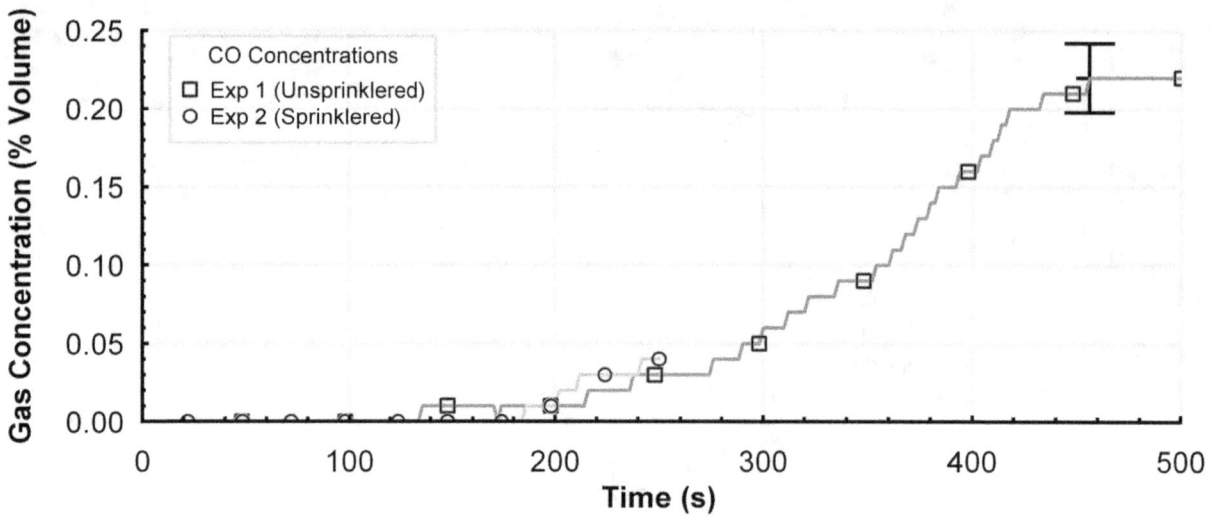

Figure 4.1-5. Carbon monoxide concentrations versus time in the dorm room from Experiment 1 and 2 at approximately 1.5 m (5 ft) above the floor.

4.2 Comparison of the Non-Sprinklered Experiments

Three of the five Experiments, 1, 4 and 5, were conducted without sprinklers. The first experiment had the dorm room door to the corridor closed, while the door was open in the latter two experiments (4 and 5). The photos and data that follows show the impact that a closed door can have on the room of origin.

Figure 4.2-1. Post fire photographs of the dorm room for Experiment 1 (door closed) on the left and Experiment 5 (door open) on the right.

Figure 4.2-2. Post fire photographs of the corridor for Experiment 1 (door closed) on the left and Experiment 5 (door open) on the right.

Figure 4.2-3 and Figure 4.2-4 show the temperature time histories in the dorm room and the corridor locations for Experiment 1, 4 and 5. In all three cases, the temperatures in the dorm room reached untenable levels in excess of 120 °C (248 °F). Figure 4.2-4 shows that the corridor temperature approached 120 °C (248 °F) in the open door experiments, while the temperature remained near ambient for the closed door experiment.

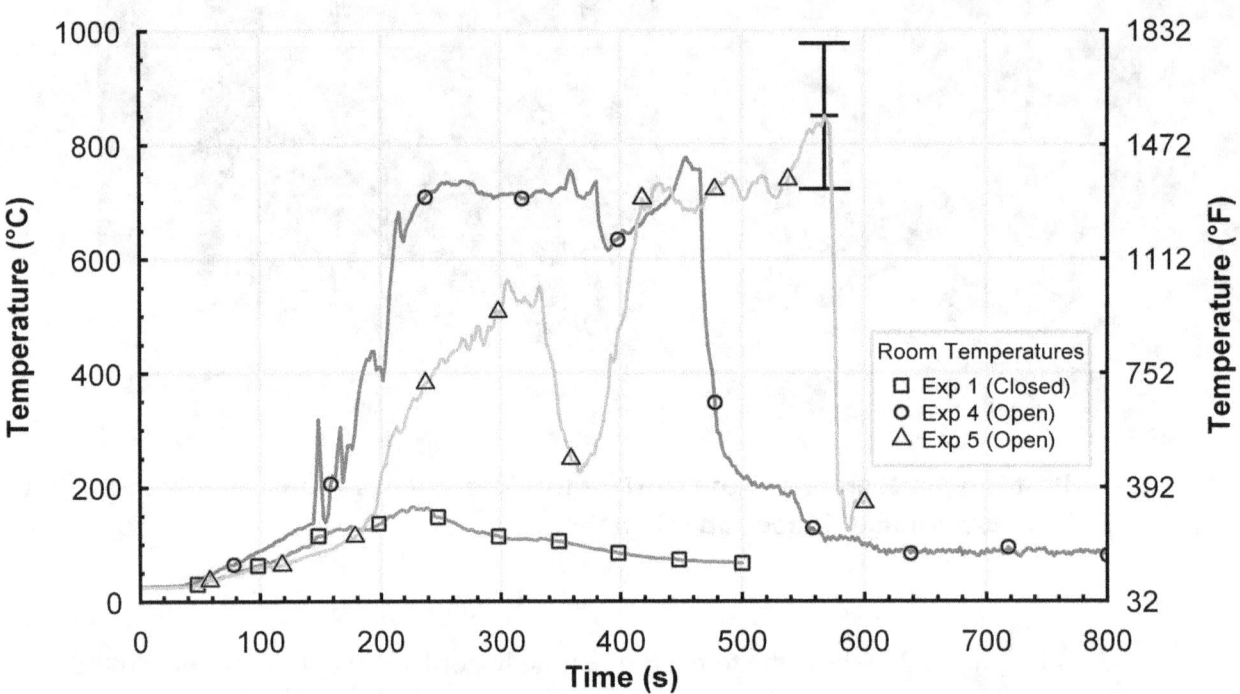

Figure 4.2-3. Temperature versus time data from thermocouple array 1 from Experiment 1, 4 and 5 at approximately 1.5 m (5 ft) above the floor.

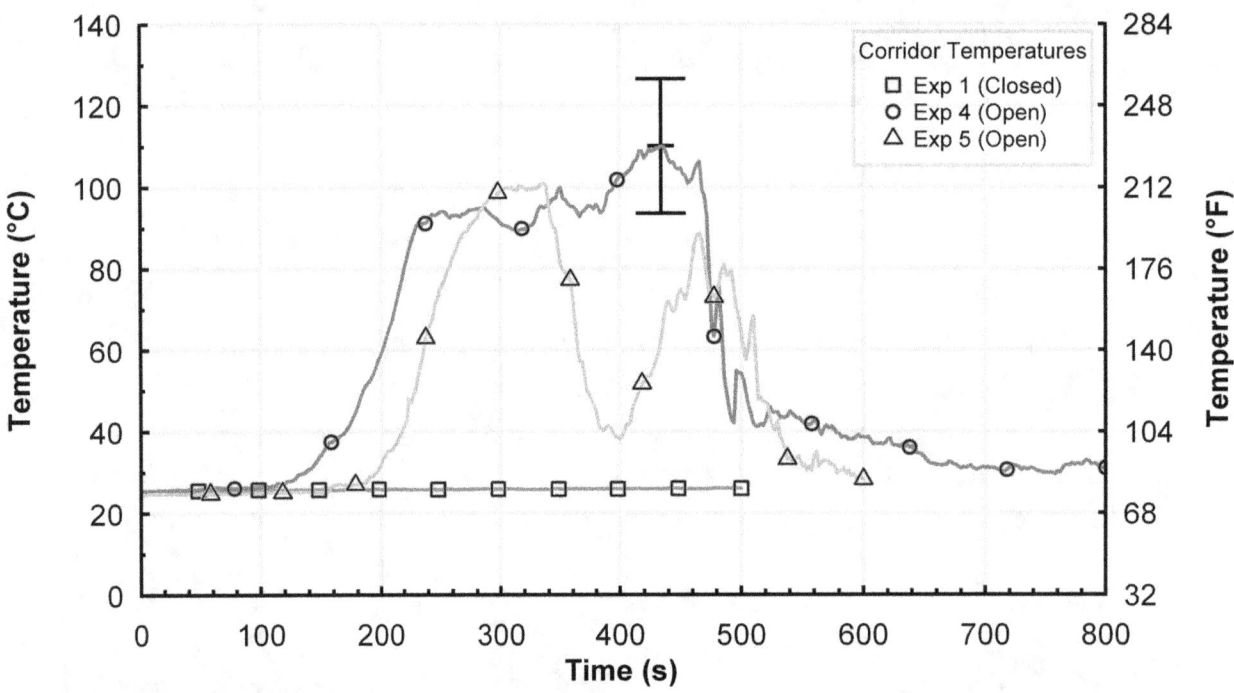

Figure 4.2-4. Temperature versus time data from thermocouple array 4 from Experiment 1, 4 and 5 at approximately 1.5 m (5 ft) above the floor.

The oxygen measurements in the dorm room and the corridor are shown in Figure 4.2-5 and Figure 4.2-6. The closed door prevented sufficient oxygen from entering the room. Therefore the fire was unable to generate additional energy and the fire self-extinguished when the oxygen level fell below 15%. In the open door experiments, the fires continued to grow until they transitioned to flashover, which in turn consumed most of the oxygen in the dorm room. In addition to generating untenable temperatures in the dorm rooms, there were untenable gas concentrations in the corridor as shown Figure 4.2-5, Figure 4.2-7 and Figure 4.2-9. This led to incapacitating conditions in the corridor due to reduced oxygen, and increased carbon dioxide and carbon monoxide as shown in Figure 4.2-6, Figure 4.2-8 and Figure 4.2-10. The closed door prevented heat or smoke from being transported to the corridor in Experiment 1.

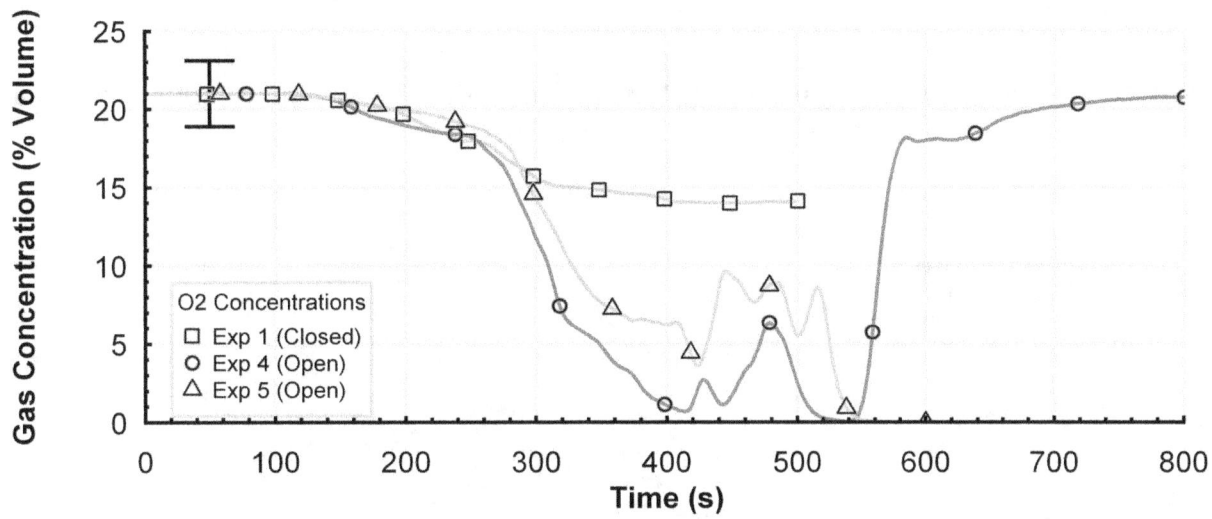

Figure 4.2-5. Oxygen concentrations versus time in the dorm room from Experiment 1, 4 and 5 at approximately 1.5 m (5 ft) above the floor.

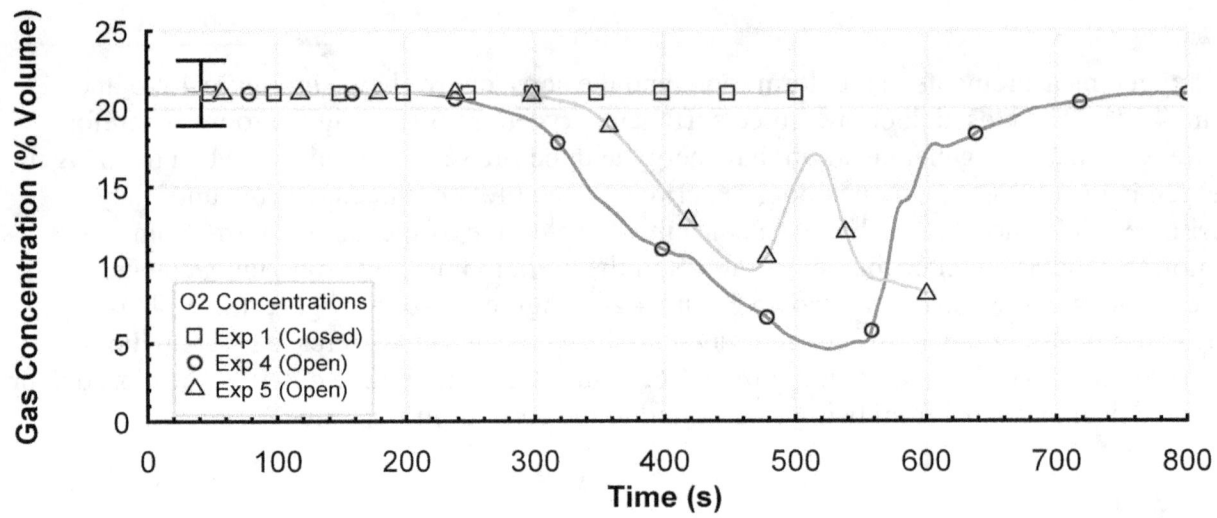

Figure 4.2-6. Oxygen concentrations versus time in the corridor from Experiment 1, 4 and 5 at approximately 1.5 m (5 ft) above the floor.

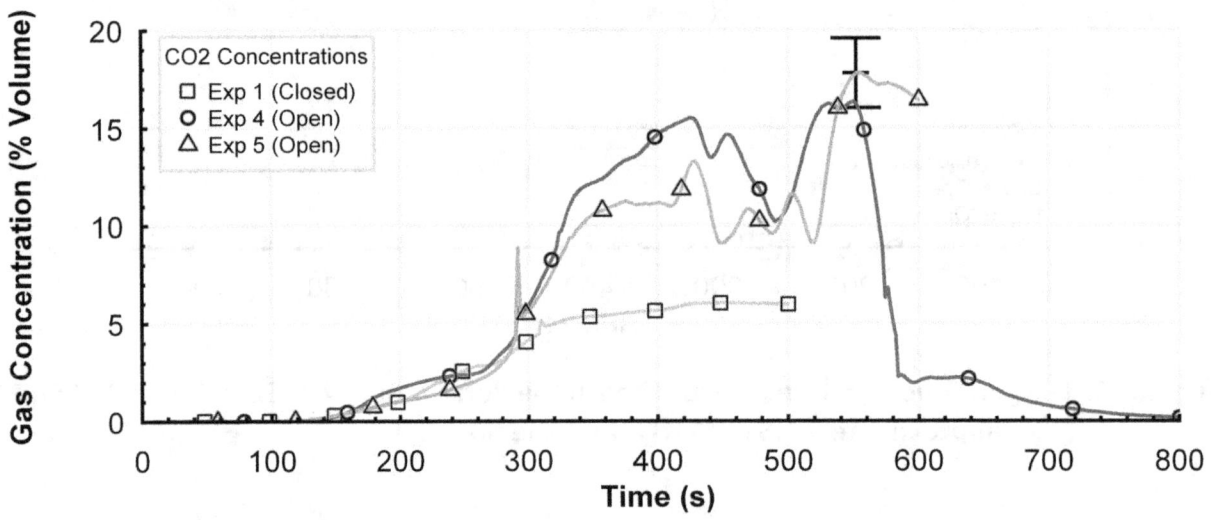

Figure 4.2-7. Carbon dioxide concentrations versus time in the dorm room from Experiment 1, 4 and 5 at approximately 1.5 m (5 ft) above the floor.

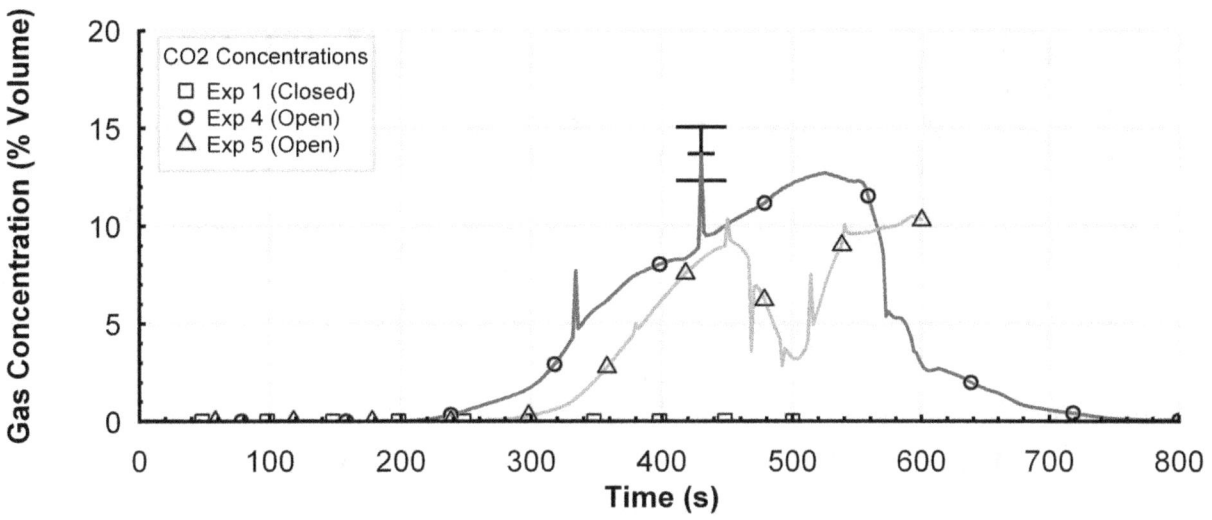

Figure 4.2-8. Carbon dioxide concentrations versus time in the corridor from Experiment 1, 4 and 5 at approximately 1.5 m (5 ft) above the floor.

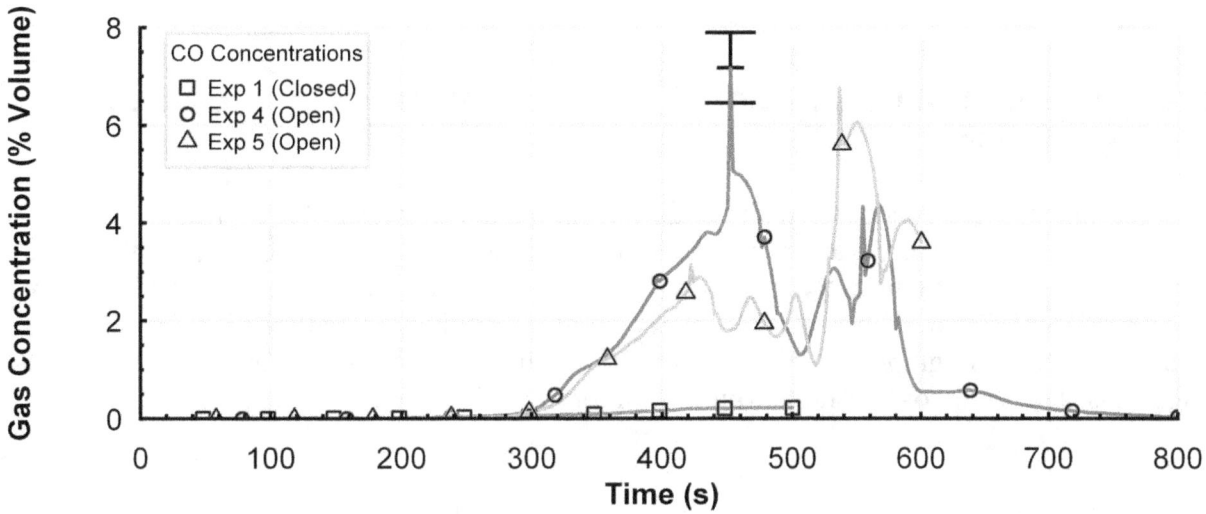

Figure 4.2-9. Carbon monoxide concentrations versus time in the dorm room from Experiment 1, 4 and 5 at approximately 1.5 m (5 ft) above the floor.

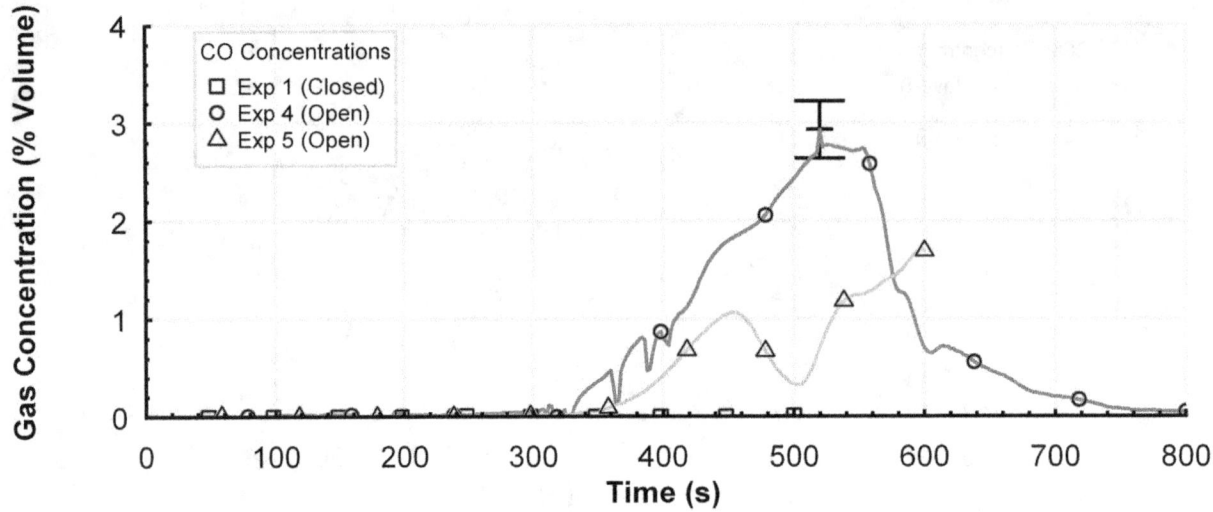

Figure 4.2-10. Carbon monoxide concentrations versus time in the corridor from Experiment 1, 4 and 5 at approximately 1.5 m (5 ft) above the floor.

4.3 Comparison of Open Door, Sprinklered versus Non-sprinklered Experiments

Three experiments were conducted with the door from the dorm room to the corridor open. In one case, Experiment 3, there was an automatic fire sprinkler in the dorm room and in the other two experiments (4 and 5), no active sprinkler was provided. The photographs shown in Figure 4.3-1 are representative of the results. With an operating sprinkler present, the fire damage was very limited. The photograph of the dorm room taken after Experiment 4 shows a room that has damage consistent with post-flashover (burning floor to ceiling) fire.

Figure 4.3-1. Post fire photographs for Experiment 3 (sprinklered) on the left and Experiment 4 (non-sprinklered) on the right.

The comparative temperatures from the dorm room and corridor are shown in Figure 4.3-2 and Figure 4.3-3. In the dorm room with a sprinkler, temperatures remained tenable. As previously noted, the temperatures in the non-sprinklered dorm rooms became untenable within 182 s of ignition.

Figure 4.3-4 through Figure 4.3-9 provides comparisons of the measured gases in the dorm room and the corridor from Experiment 3, 4 and 5. The conditions in the sprinklered experiment (3) remained tenable with respect to the levels of oxygen, carbon dioxide and carbon monoxide that were present in both the dorm room and the corridor. The two non-sprinklered experiments transitioned to flashover. In both experiments, the levels of oxygen, carbon dioxide and carbon monoxide were untenable, as defined earlier in this section, within 346 s and 454 s after ignition for the dorm room and the corridor respectively, The time to reach the untenability criteria for each gas is presented in Table 4.4-2.

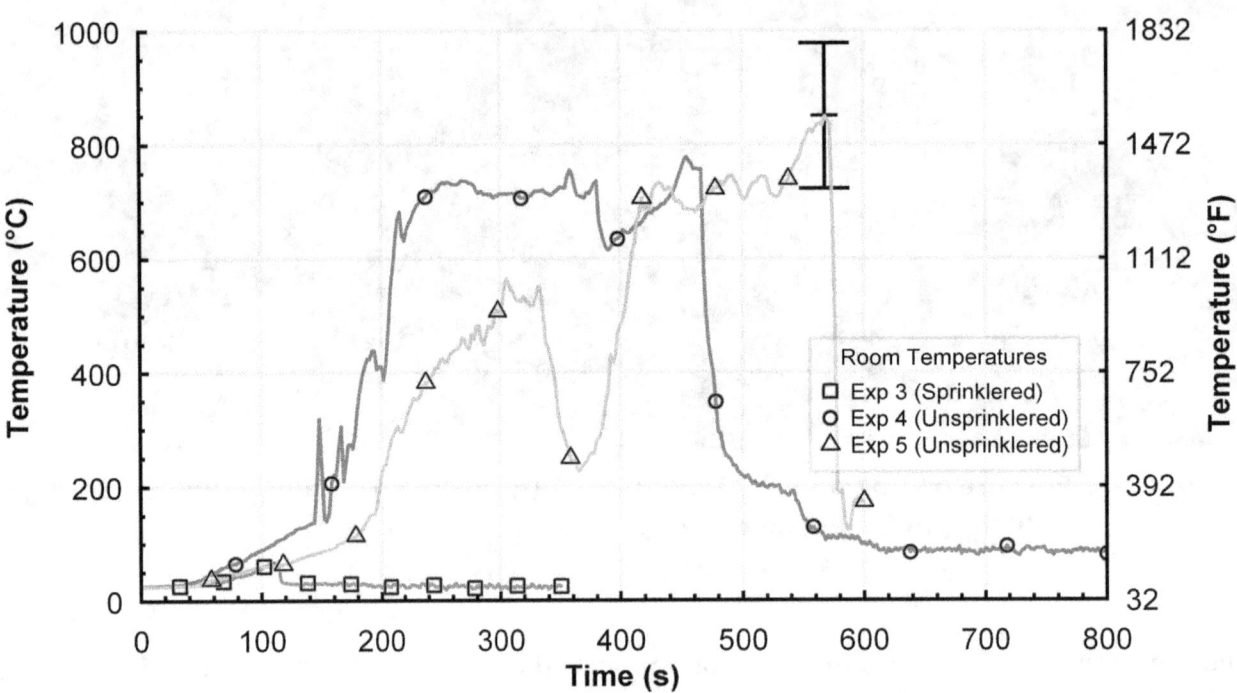

Figure 4.3-2. Temperature versus time data from thermocouple array 1 from Experiment 3, 4 and 5 at approximately 1.5 m (5 ft) above the floor.

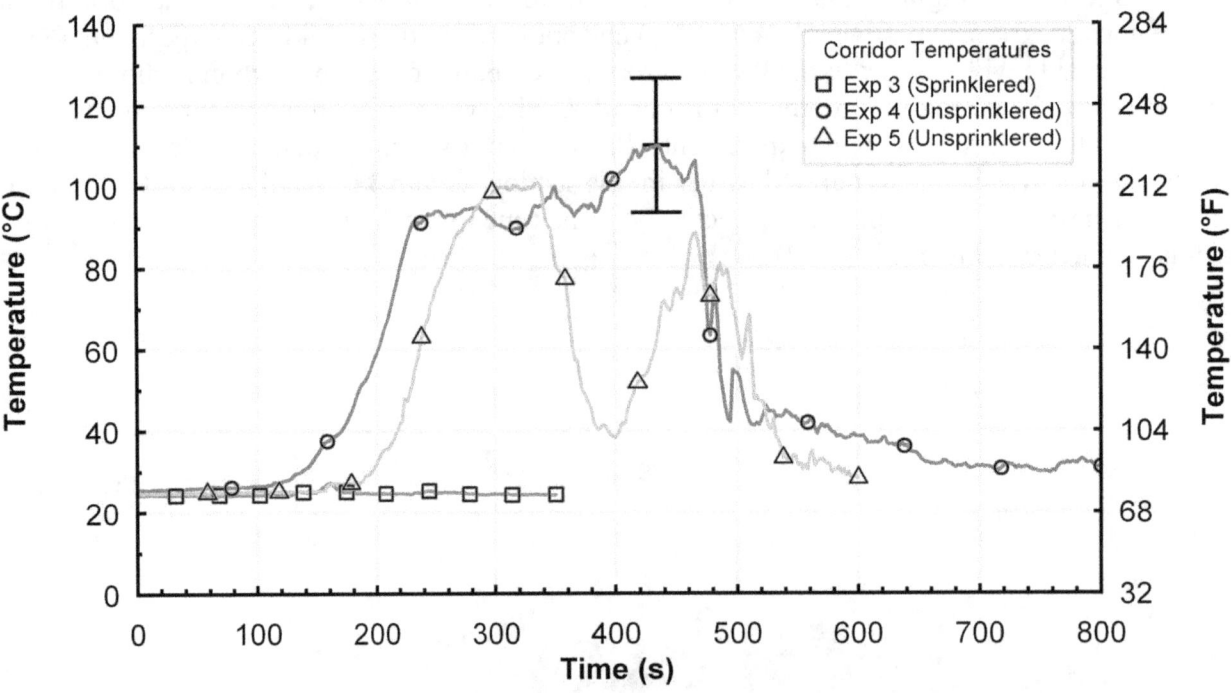

Figure 4.3-3. Temperature versus time data from thermocouple array 4 from Experiment 3, 4 and 5 at approximately 1.5 m (5 ft) above the floor.

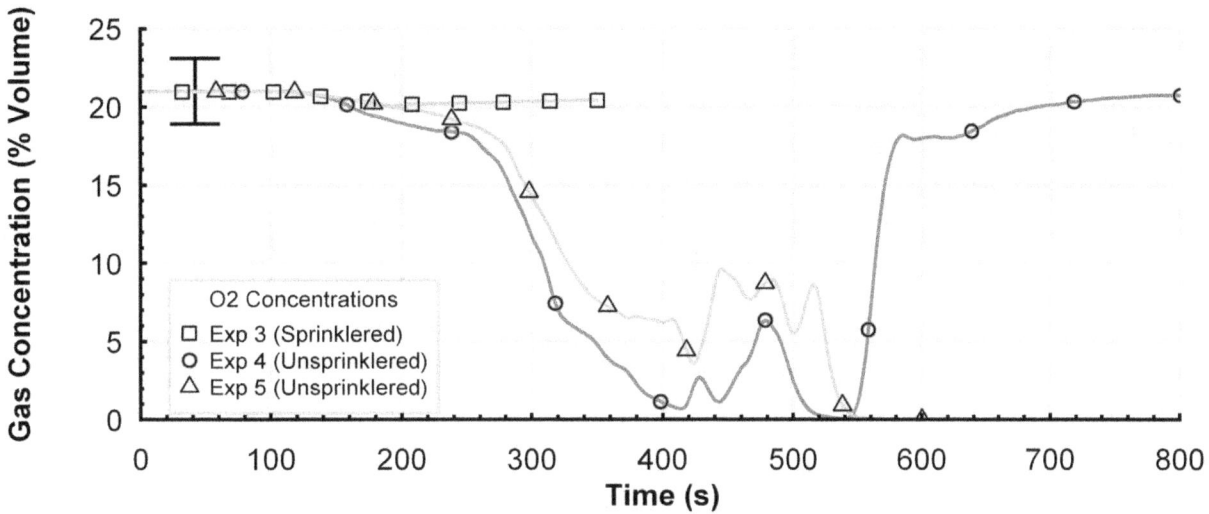

Figure 4.3-4. Oxygen concentrations versus time in the dorm room from Experiment 3, 4 and 5 at approximately 1.5 m (5 ft) above the floor.

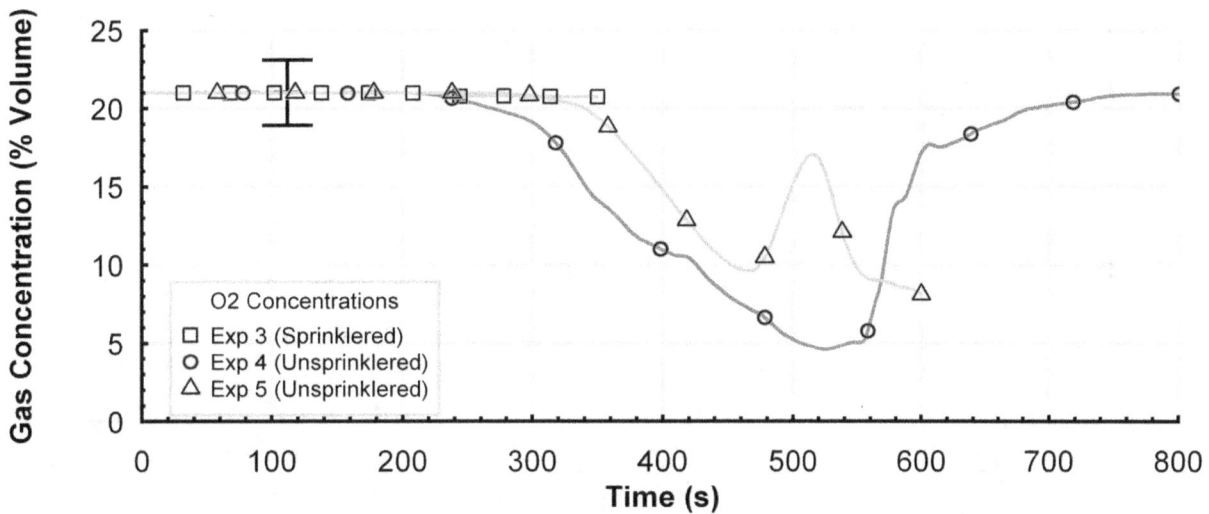

Figure 4.3-5. Oxygen concentrations versus time in the corridor from Experiment 3, 4 and 5 at approximately 1.5 m (5 ft) above the floor.

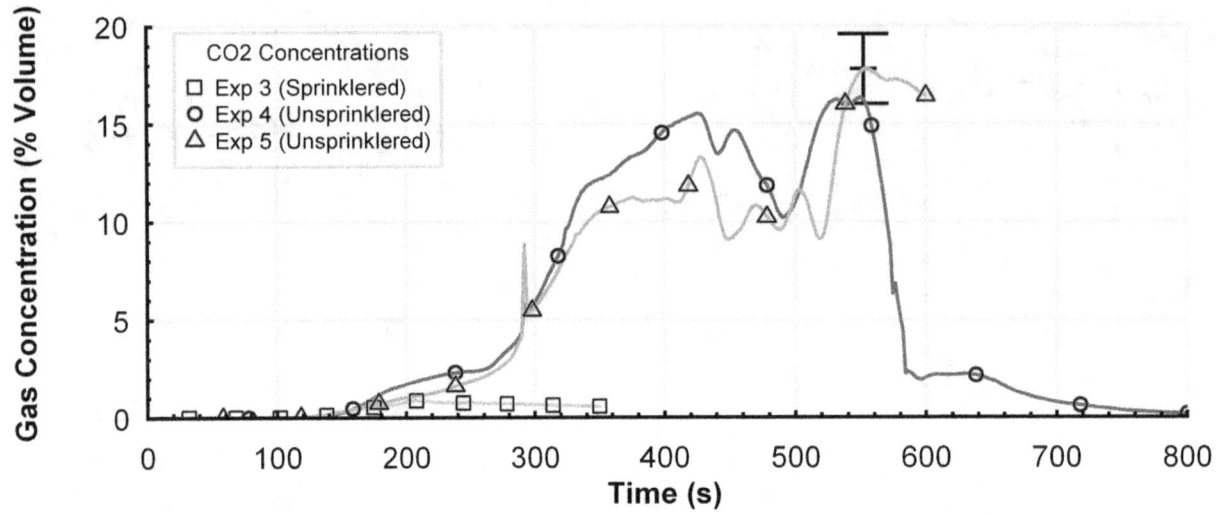

Figure 4.3-6. Carbon dioxide concentrations versus time in the dorm room from Experiment 3, 4 and 5 at approximately 1.5 m (5 ft) above the floor.

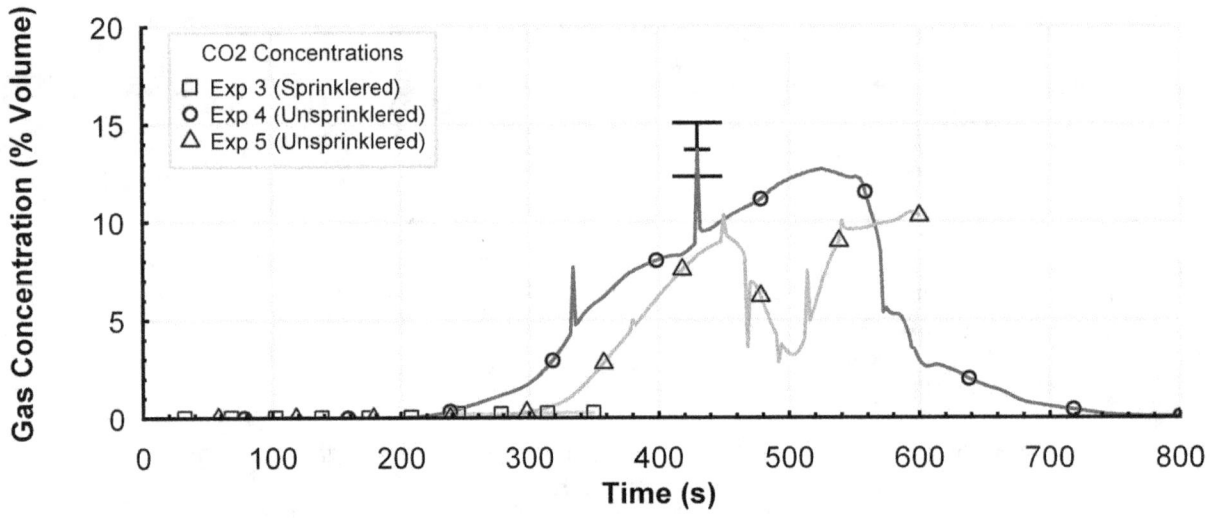

Figure 4.3-7. Carbon dioxide concentrations versus time in the corridor from Experiment 3, 4 and 5 at approximately 1.5 m (5 ft) above the floor.

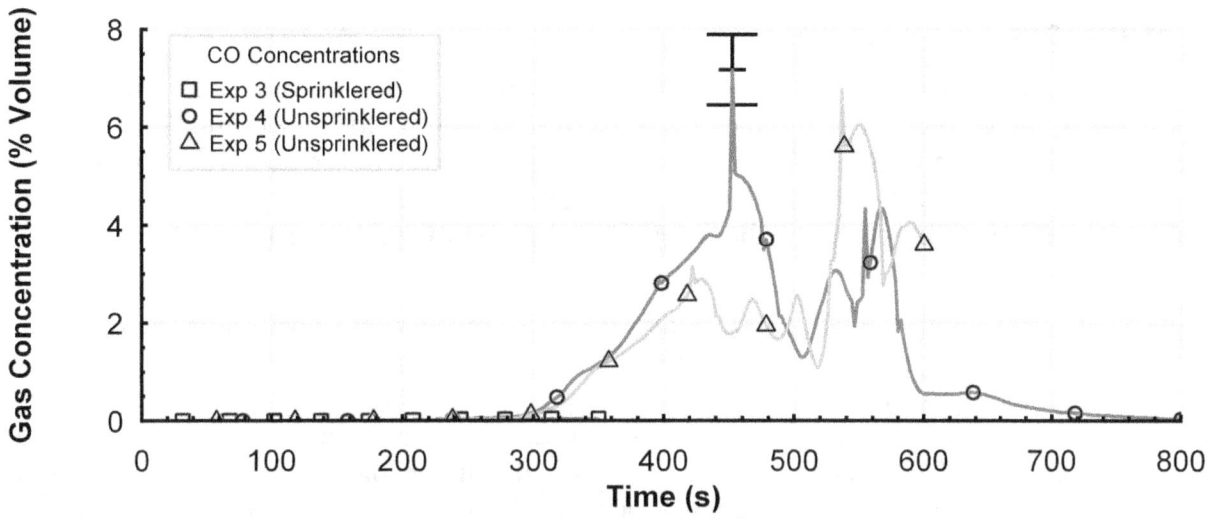

Figure 4.3-8. Carbon monoxide concentrations versus time in the dorm room from Experiment 3, 4 and 5 at approximately 1.5 m (5 ft) above the floor.

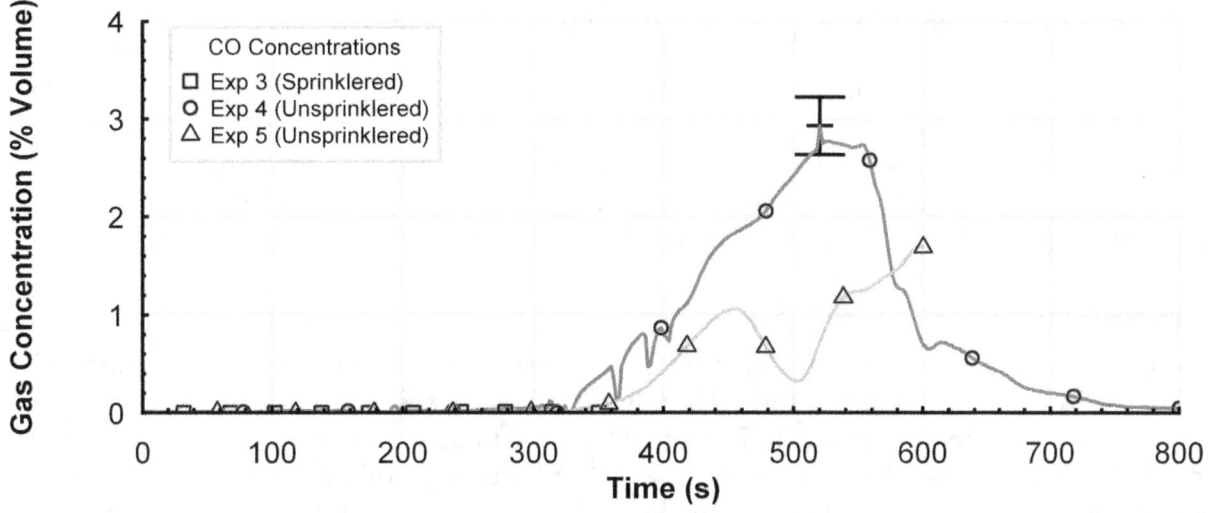

Figure 4.3-9. Carbon monoxide concentrations versus time in the corridor from Experiment 3, 4 and 5 at approximately 1.5 m (5 ft) above the floor.

4.4 Summary

The complete set of measurements from each of the five experiments was presented in Section 3. In the previous portions of this section comparisons have been provided to examine the impact of automatic sprinkler protection and the impact of keeping the door to the room of origin closed. In this summary section, Table 4.4-1 presents the smoke alarm and sprinkler activation times for comparison against the times to reach the untenability criteria in the dorm room and the corridor, given in Table 4.4-2.

In each of the experiments, the ionization smoke alarms installed in the dorm rooms activated within 12 s to 26 s after ignition. The average time between the first smoke alarm activation and the time to reach untenability in the dorm room for the three unsprinklered experiments (Experiment 1, 4 and 5) was 134 s. The last two experiments (Experiment 4 and 5), which were unsprinklered and had the dorm room door open, resulted in untenable conditions in the corridor as well as in the dorm room. The average time between the activation of smoke alarm in the center location in the corridor and the time to reach untenable conditions in the corridor was 356 s.

The untenable conditions in the corridor for Experiment 4 and 5, would represent the worst case for a building occupant, not located in the room of origin, who needed to use the corridor as a means of egress. In these experiments, the warning of the smoke alarms located in the corridor provided at least 5 minutes of available safe egress time.

Experiment 1 and 2 were conducted with the dorm room door closed. In both cases the corridor remained tenable throughout the duration of the experiments. In Experiment 1, which was unsprinklered, the closed door limited the availability of fresh oxygen to the fire, which resulted in the limitation of the amount of heat that could be released and eventually led to the self-extinguishment of the fire. While the thermal tenability limit was exceeded in the dorm room, the transmission of hazardous conditions from the fire to building occupants outside the room of origin were mitigated by the closed door.

Experiment 2 and 3 had active sprinkler systems installed. The time between the activations of the smoke alarm and the automatic sprinkler in the same room was approximately 90 s in both experiments. Experiment 3 had the dorm room door open to the corridor. As a result, the smoke alarms in the corridor also activated prior to the thermal activation of the sprinkler in the room. In both of the sprinklered experiments (Experiment 2 and 3), the tenability limits were not exceeded.

Table 4.4-1. Sprinkler and Smoke Alarm Activation Times and Temperatures at Those Devices at the Time of Activation

Experiment	Door Position	Active Sprinkler System	Room Smoke Alarm Activation Time/Temp (s) / (°C)	West Corridor Smoke Alarm Activation Time/Temp (s) / (°C)	Center Corridor Smoke Alarm Activation Time/Temp (s) / (°C)	East Corridor Smoke Alarm Activation Time/Temp (s) / (°C)	Room Sprinkler Activation Time/Temp (s) / (°C)	Corridor Sprinkler Activation Time/Temp (s) / (°C)
1	Closed	No	24 / 52	160 / 27	216 / 27	316 / 27	120 / 118	NA
2	Closed	Yes	12 / 32	NA	NA	NA	105 / 119	NA
3	Open	Yes	22 / 46	68 / 27	36 / 28	62 / 29	112 / 112	NA
4	Open	No	14 / 45	60 / 31	32 / 29	62 / 32	76 / 136	128 / 99
5	Open	No	26 / 31	62 / 30	80 / 31	98 / 31	110 / 82	224 / 125

Table 4.4-2. Time to Reach Given Untenability Criteria (Red) or Most Significant Tenability Risk Encountered (Blue) for the Dorm Room and Corridor

Experiment	Door Position	Active Sprinkler System	Dorm Room				Corridor			
			Temp. > 120 °C	O_2 < 10 %	CO_2 > 8 %	CO > 1 %	Temp. > 120 °C	O_2 < 10 %	CO_2 > 8 %	CO > 1 %
1	Closed	No	156 s	14 %	6 %	0 %	26 °C	21 %	0 %	0 %
2	Closed	Yes	84 °C	20 %	1 %	0 %	27 °C	21 %	0 %	0 %
3	Open	Yes	68 °C	20 %	1 %	0 %	27 °C	21 %	0 %	0 %
4	Open	No	128 s	310 s	318 s	340 s	110 °C	426 s	398 s	410 s
5	Open	No	182 s	328 s	292 s	346 s	101 °C	454 s	426 s	444 s

Experiment 4 and 5 also showed the impact of ventilation on fire growth and development. In Experiment 4, the first window pane completely vented open at 220 s after ignition, which coincided with flashover. By 328 s after ignition, all three window panes were gone and flames were venting out of the openings. In Experiment 5, the fire development was slower than that in Experiment 4. The first window pane completely vented open at 331 s after ignition. It took another 95 s for the other two window panes to self-vent, providing the dorm room with the ventilation needed to achieve flashover. The fire behavior in Experiment 5 transitioned from a fuel limited fire to an under-ventilated, fuel-rich condition, which resulted in reduced temperatures in the dorm room. Once all of the window openings were completely vented open, enough fresh oxygen could be drawn into the room to increase the heat release rate, resulting in increased temperatures in the room and leading to flashover at approximately 440 s after ignition, see Figure 3.5-1 and Figure 3.5-2.

The conditions in the corridor during Experiment 4 and 5 also provide insight into the conditions that firefighters would face under near ideal natural ventilation conditions with ambient air being introduced from the ends of the corridor while the flames and majority of the hot gases were exhausted out of the open windows of the dorm room. The peak thermal conditions in the corridor during two experiments, at approximately 0.9 m (3 ft) above the floor, were approximately 75 °C (170 °F), with a total heat flux of approximately 2.5 kW/m^2 (0.22 BTU/ft^2·s).

5 Conclusions

This report describes a series of experiments in which fires were initiated in dormitory sleeping rooms. These experiments were conducted by NIST in cooperation with the University of Arkansas and the Fayetteville Fire Department with the support of the U.S. Fire Administration.

The experimental conditions were documented, including a description of the building geometry and construction, the fuel load in the dorm rooms, and the location of the instrumentation used to measure gas temperature, oxygen, carbon dioxide and carbon monoxide concentrations, and heat flux. Smoke alarm activation and sprinkler activation times were also reported. Five experiments were conducted. In two of the experiments, the door between the dorm room (room of fire origin) and the corridor was closed. In the other three experiments, the door from the dorm room (room of fire origin) remained open to the corridor. In each case, door closed or door open, one of the experiments included an operating automatic fire sprinkler. The results from the experiments comparing the sprinklered and non-sprinklered dorm rooms were presented, as well as a comparison of the results from the open and closed corridor door experiments.

In each of the experiments the smoke alarm activated prior to the development of untenable conditions in the room of origin, within 12 s to 26 s after ignition. In all of the experiments, the smoke alarm sounded while the fire was limited to the first fuel ignited.

Experiment 1 and 2 were conducted with the dorm room door closed. In both of the closed door experiments, the corridor remained tenable throughout the duration of the experiments.

Experiment 2 and 3 had active automatic fire sprinklers installed. The time between the activation of the smoke alarm located in the dorm room and the activation of the automatic sprinkler in the room was approximately 90 s in both experiments. In the sprinklered experiments, tenability was maintained in the dorm room and the corridor.

Experiments 4 and 5 were conducted with the dorm room door open and no active sprinkler. In both of these experiments, the tenability limits were exceeded in the dorm room and the corridor.

The results from these experiments demonstrate the potential life safety benefits of smoke alarms for early detection and notification and automatic fire sprinkler systems for fire suppression and life safety in college dormitories and similar occupancies. The benefit of a closed door between the fire room and corridor in limiting the spread of smoke and gasses to other areas of the building was also demonstrated.

6 References

1. Flynn, Jennifer D., *Structure Fire in Dormitories, Fraternities, Sororities and Barracks*. National Fire Protection Association, Quincy. MA., August 2009.

2. Madrzykowski, D., Stroup, D.W., Walton, W.D., "Impact of Sprinklers on the Fire hazard in Dormitories: Day Room Experiments," National Institute of Standards and Technology, Gaithersburg, MD, June 2004.

3. Stroup, D.W., DeLauter, L.A., Lee, J.H., Roadarmel, G.L., "Large Fire Research Facility (Building 205) Exhaust Hood Heat Release Rate Measurement System," NISTIR 6509, National Institute of Standards and Technology, Gaithersburg, MD, July 2000.

4. Taylor, B.N. and Kuyatt, C.E., "Guidelines for Evaluating and Expressing the Uncertainty of NIST Measurement Results," NIST TN 1297, National Institute of Standards and Technology, Gaithersburg, MD, January 1994.

5. Omega Engineering Inc., *The Temperature Handbook*, Vol. MM, pages Z-39-40, Stamford, CT., 2004.

6. Blevins, L.G., "Behavior of Bare and Aspirated Thermocouples in Compartment Fires", *National Heat Transfer Conference, 33rd Proceedings*. HTD99-280. August 15-17, 1999, Albuquerque, NM, 1999.

7. Pitts, W.M., E. Braun, R.D. Peacock, H.E. Mitler, E. L. Johnsson, P.A. Reneke, and L.G.Blevins, "Temperature Uncertainties for Bare-Bead and Aspirated Thermocouple Measurements in Fire Environments," *Thermal Measurements: The Foundation of Fire Standards. American Society for Testing and Materials (ASTM). Proceedings*. ASTM STP 1427. December 3, 2001, Dallas, TX.

8. Medtherm Corporation Bulletin 118, "64 Series Heat Flux Transducers", Medtherm Corporation, Huntsville, AL., August 2003.

9. Pitts, William, M., Annageri V. Murthy, John L. de Ris, Jean-Rémy Filtz, , Kjell Nygard, Debbie Smith, and Ingrid Wetterlund. *Round robin study of total flux gauge calibration at fire laboratories,* Fire Safety Journal 41, 2006, pp 459-475.

10. Bundy, M., Hamins, A., Johnsson, E.L., Kim, S.C., Ko, G.H., and Lenhart, D. B., "Measurements of Heat and Combustion Products in Reduced-Scale Ventilated-Limited Compartment Fires", National Institute of Standards and Technology, Gaithersburg, MD., NIST TN 1483, July 2007.

11. Omega Engineering Inc., Flow, Level, and Environmental Handbook Omega Engineering Inc., Stamford, CT 2000

12. Ohaus Corporation, Manual for SD Series Bench Scale, Pine Brook, NJ., 2000.

13. Purser, David A., "Assessment of Hazards to Occupants from Smoke, Toxic Gases, and Heat," *SFPE Handbook of Fire Protection Engineering, 4th ed.* DiNenno, P. J., Drysdale, D., Beyler, C.L., Walton, W. D., Custer, R.L.P., Hall J.R., Watts, J.M. (eds). NFPA Quincy, MA, 2008.

www.ingramcontent.com/pod-product-compliance
Lightning Source LLC
Chambersburg PA
CBHW081826170526
45167CB00007B/2732